W9-BYT-234

How To Solve Word Problems in Algebra

A Solved Problem Approach

How To Solve Word Problems in Algebra

A Solved Problem Approach

by

MILDRED JOHNSON

Professor Emeritus of Mathematics
Chaffey Community College
Alta Loma, California

McGRAW-HILL, INC.
New York St. Louis San Francisco Auckland Bogotá
Caracas Lisbon London Madrid Mexico Milan Montreal
New Delhi Paris San Juan Singapore Sydney Tokyo Toronto

Revised printing, 1992
Copyright © 1976 by McGraw-Hill, Inc. All rights reserved. Printed in the United States of America. Except as permitted under the United States Copyright Act of 1976, no part of this publication may be reproduced or distributed in any form or by any means, or stored in a data base or retrieval system, without the prior written permission of the publisher.

20 FGRFGR 9 9 8 7 6 5 4

ISBN 0-07-032631-2 (Formerly published under ISBN 0-07-032620-7.)

Library of Congress Cataloging-in-Publication Data

Johnson, Mildred (date).
How to solve word problems in algebra: a solved problem approach / Mildred Johnson. —updated 1st ed.
p. cm.
ISBN 0-07-032631-2
1. Algebra—Problems, exercises, etc. 2. Problem solving.
I. Title.
QA157.J7 1994
512'.9'0076—dc20 93-11236

Preface

There is no area in algebra which causes students as much difficulty as word problems. Most textbooks in algebra do not have adequate explanations and examples for the student who is having trouble with them. This book's purpose is to give the student detailed instructions in procedures and many completely worked examples to follow. All major types of word problems usually found in algebra texts are here. Emphasis is on the mechanics of word-problem solving because it has been my experience that students having difficulty can learn basic procedures even if they are unable to reason out a problem.

This book may be used independently or in conjunction with a text to improve skills in solving word problems. The problems are suitable for either elementary or intermediate level algebra students. A supplementary miscellaneous problem set with answers only is at the end of the book, for drill or testing purposes.

MILDRED JOHNSON

Contents

Introduction

How To Work Word Problems

If you are having trouble with word problems, this book is for you! In it you will find many examples of the basic types of word problems completely worked out for you. Learning to work problems is like learning to play the piano. First you are shown how. Then you must practice and practice and practice. Just *reading* this book will not help unless you *work* the problems. The more you work, the more confident you will become. After you have worked many problems with the solutions there to guide you, you will find miscellaneous problems at the end of the book for extra practice.

You will find certain basic types of word problems in almost every algebra book. You can't go out and use them in daily life, or in electronics, or in nursing. But they teach you *basic procedures* which you will be able to use elsewhere. This book will show you step-by-step what to do in each type of problem. Let's learn how it's done!

How do you start to work a word problem?

1. *Read the problem* all the way through quickly, to see what kind of word problem it is and what it is about.
2. Look for a question at the end of the problem. This is often a good way to find what you are solving for. Sometimes two or three things need to be found.
3. Start every problem with "Let x = something." (We generally use x for the unknown.) You let x equal what you are trying to find. What you are trying to find is usually stated in the question at the end of the problem. This is called the *unknown.* You must show and label what x stands for in your problem or your equation has no meaning. You'll note in each solved problem in this book

that x is always labeled with the unit of measure called for in the problem (inches, mph, pounds, etc.). That's why we don't bother repeating the units label for the answer line.

4. If you have to find more than one quantity or unknown, try to determine the *smallest* unknown. This unknown is often the one to let x equal.
5. Go back and read the problem over again. This time read it a piece at a time. Simple problems generally have two statements. One statement helps you set up the unknowns, and the other gives you equation information. Translate the problem from words to symbols a piece at a time.

Here are some examples of statements translated into algebraic language, using x as the unknown. From time to time refer to these examples to refresh your memory as you work the problems in this book.

Statement	Algebra
1. *Twice as much as* the unknown	$2x$
2. *Two less than* the unknown	$x-2$
3. *Five more than* the unknown	$x+5$
4. *Three more than twice* the unknown	$2x+3$
5. A number decreased by 7	$x-7$
6. Ten decreased by the unknown	$10-x$
7. Sheri's age (x) 4 years from now	$x+4$
8. Dan's age (x) 10 years ago	$x-10$
9. Number of cents in x quarters	$25x$
10. Number of cents in $2x$ dimes	$10(2x)$
11. Number of cents in $x+5$ nickels	$5(x+5)$
12. Separate 17 into two parts	x and $17-x$
13. Distance traveled in x hours at 50 mph	$50x$
14. Two consecutive integers	x and $x+1$
15. Two consecutive even integers	x and $x+2$

16.	Two consecutive odd integers	x and $x+2$
17.	Interest on x dollars for 1 year at 5%	$0.05x$
18.	\$20,000 separated into two investments	x and $\$20,000-x$
19.	Distance traveled in 3 hours at x mph	$3x$
20.	Distance traveled in 40 minutes at x mph (40 minutes = $\frac{2}{3}$ hours)	$\frac{2x}{3}$
21.	Sum of a number and 20	$x+20$
22.	Product of a number and 3	$3x$
23.	Quotient of a number and 8	$\frac{x}{8}$
24.	Four times as much	$4x$
25.	Three is four more than a number	$3=x+4$

FACTS TO REMEMBER

1. *Times as much* means multiply.
2. *More than* means add.
3. *Decreased by* means subtract.
4. *Increased by* means add.
5. Separate 28 into two parts means find two numbers whose sum is 28.
6. *Percent of* means multiply.
7. *Is, was, will be,* become the equals sign ($=$) in algebra.
8. If 7 exceeds 2 by 5, then $7-2=5$. *Exceeds* becomes a minus sign ($-$) and *by* becomes an equals sign ($=$).
9. No unit labels such as feet, degrees, and dollars are used in equations. In this book we have left these labels off the answers as well. Just refer to the "Let $x =$" statement to find the unit label for the answer.

EXAMPLES OF HOW TO START A PROBLEM

1. One number is two times another.

Let x = smaller number

$2x$ = larger number

2. A man is 3 years older than twice his son's age.

Let x = son's age

$2x + 3$ = man's age

3. Represent two numbers whose sum is 72.

Let x = one number

$72 - x$ = the other number

4. A man invested \$10,000, part at 5% and part at 7%. Represent interest (income).

Let x = amount invested at 5%

$\$10,000 - x$ = amount invested at 7%

Then $0.05x$ = interest on 1st investment

$0.07(10,000 - x)$ = interest on 2nd investment

When money is invested, rate of interest times principal equals amount of interest per year.

5. A mixture contains 5% sulphuric acid. Represent the amount of acid (in quarts).

Let x = number of quarts in the mixture

$0.05x$ = number of quarts of sulphuric acid

6. A woman drove for 5 hours at a uniform rate per hour. Represent her distance traveled.

Let x = rate in miles per hour

$5x$ = number of miles traveled

7. A girl had 2 more dimes than nickels. Represent how much money she had in cents.

$$
\begin{aligned}
\text{Let} \qquad x &= \text{number of nickels} \\
x + 2 &= \text{number of dimes} \\
5x &= \text{number of } \textit{cents} \text{ in nickels} \\
&\quad \text{(5 cents in } \textit{each} \text{ nickel)} \\
10(x+2) &= \text{number of cents in dimes} \\
&\quad \text{(10 cents in each dime)}
\end{aligned}
$$

FACTS TO REMEMBER ABOUT SOLVING AN EQUATION

(These are facts you should already have learned about procedures in problem solutions.)

1. Remove parentheses first.

   subtraction $-(3x+2) = -3x-2$
   multiplication (distributive law) $3(x+2) = 3x+6$

2. Remove fractions by multiplying by lowest common denominator.

   $$\frac{3}{x} + 4 = \frac{5}{2x} + 3$$

   The LCD is $2x$. Multiplying both sides of the equation by $2x$,

   $$6 + 8x = 5 + 6x$$

3. Decimals should be removed from an equation before solving. Multiply by a power of 10 large enough to make all decimal numbers whole numbers. If you multiply by 10, you move the decimal point in all terms *one* place to the right. If you multiply by 100 you move the decimal point in all terms *two* places to the right.

   Example

   $$0.03x + 201.2 - x = 85$$

   Multiply both sides of the equation by 100.

   $$3x + 20{,}120 - 100x = 8500$$

Now let's try some word problems. In this book they are grouped into general types. That way you can look up each kind of problem for basic steps if you are using an algebra text.

Chapter 1

Numbers

Number problems are problems about the relationships among numbers. The unknown is a whole number, not a fraction or a mixed number. It is almost always a positive number. In some problems the numbers are referred to as *integers*, which you may remember are positive numbers, negative numbers, or zero.

EXAMPLE 1. There are two numbers whose sum is 72. One number is twice the other. What are the numbers?

Steps

1. Read the problem. It is about numbers.

2. The question at the end asks, "What are the numbers?" So we start out "let x = smaller number." Be sure you always start with x (that is, $1x$). Never start off with "let $2x$ = something," because it doesn't have any meaning unless you know what x stands for. Always label x as carefully as you can. In this problem, that means just the label "smaller number," since no units of measure are used in this problem.

3. Read the problem again, a piece at a time. The first line says you have two numbers. So far, you have one number represented by x. You know you have to represent two unknowns because the problem asks you to find two numbers. So read on. Next the problem states that the sum of the two numbers equals 72. Most of the time it is good to save a sum for the equation statement if you can. A sum may also be used to represent the second unknown, as you shall see later. The next statement says *one number is twice another*. Here is a fact for the second unknown!

Now you have

$$\text{Let} \qquad x = \text{smaller number}$$
$$2x = \text{larger number}$$

4. Now that both unknowns are represented, we can set up the equation with the fact which has not been used. *The sum is 72* has not been used; translating it, we get this equation

$$x + 2x = 72$$

Now let's put it all together and solve.

Solution

$$\text{Let} \qquad x = \text{smaller number}$$
$$2x = \text{larger number}$$

$$\text{Then} \qquad x + 2x = 72$$
$$3x = 72$$
$$\left.\begin{aligned} x &= 24 \\ 2x &= 48 \end{aligned}\right\} \quad \textit{Answers}$$

Check The sum of the numbers is 72. Thus $24 + 48 = 72$.

Be sure you have answered the question completely; that is, be sure you have solved for the unknown or unknowns asked for in the problem.

EXAMPLE 2. There are two numbers whose sum is 50. Three times the first is 5 more than twice the second. What are the numbers?

Steps

1. First carefully read the problem.

2. Reread the question at the end of the problem: "What are the numbers?"

3. Note the individual facts about the numbers:

   (*a*) There are two numbers.

   (*b*) Their sum or total is 50.

   (*c*) Three times the first is 5 more than two times the second.

This problem is different, you say. It sure is! Since statement (c) tells you to do something with the numbers, you have to represent them first. So, let's see another way to use a total or sum.

Let x = smaller number

$50 - x$ = larger number

The use of *total minus* x to represent a second unknown occurs quite often in word problems. It is like saying, "The sum of two numbers is 10. One number is 6. What is the other number?" You know that the other number is 4 because you said to yourself $10 - 6 = 4$. That is why if the sum is 50 and one number is x, we subtract x from 50 and let $50 - x$ represent the second number or unknown. If you know the sum (total), you can subtract one part (x) to get the other part.

4. Let's get back to the problem.

   (a) *two numbers whose sum is 50*

   Let x = first number

   $50 - x$ = second number

   (b) *three times the first* can be written as

   $$3x$$

   (c) *is* can be written as

   $$=$$

   (Remember that *is*, *was*, and *will be* are the *equals* sign.)

   (d) *5 more than* is

   $$+5$$

   (e) *twice the second* is

   $$2(50 - x)$$

5. Now put it together:

   Equation: $3x = +5 + 2(50 - x)$

   Only it is better to say

   $$3x = 2(50 - x) + 5$$

When you have a problem stating "five less than," write the -5 on the right of the expression. For example, five less than x translates as $x-5$. It's good to be consistent and put the "more than" and "less than" quantities on the right: $x+5$ and $x-5$; not $+5+x$ and $-5+x$.

Solution

$$3x = 2(50-x)+5$$
$$3x = 100-2x+5$$
$$5x = 105$$
$$\left.\begin{aligned} x &= 21 \\ 50-x &= 29 \end{aligned}\right\} \quad \textit{Answers}$$

Check The sum of the numbers is 50. Hence $21+29=50$ and substituting x in the equation $3x=2(50-x)+5$,

$$3(21) = 2(50-21)+5$$
$$63 = 100-42+5$$
$$63 = 63$$

EXAMPLE 3. Separate 71 into two parts such that one part exceeds the other by 7. (*Hint*: separate 71 into two parts means two numbers have a sum of 71. *Such that* means in order to have or so that.)

Steps

1. Read it.
2. What is the question (what are you looking for)? The problem really tells you to find two numbers whose sum is 71.
3. Separate 71 into two parts.

   Let
   x = smaller part
   $71-x$ = larger part (this is the *total minus* x *concept again*)
4. One part exceeds the other by seven. The larger minus the smaller equals the difference and the larger exceeds the smaller by seven.

$$(71-x)-x = 7$$

(*Exceeds* translates as minus sign and *by* as equals sign.)

Solution

$$
\begin{aligned}
(71 - x) - x &= 7 \\
71 - 2x &= 7 \\
-2x &= 7 - 71 \\
-2x &= -64 \\
x &= 32 \\
71 - x &= 39
\end{aligned}
\quad \Bigg\} \text{ Answers}
$$

Now someone may say, "But I did it differently and got the same answer." There are many different ways to do *some* problems, all of them correct. If we showed you all of them, you would be utterly confused, so if your method *always* works, you are probably correct. It is like saying:

$$
\left.
\begin{aligned}
8 - 6 &= 2 \\
8 &= 2 + 6 \\
8 - 2 &= 6
\end{aligned}
\right\} \text{ All three are correct statements}
$$

We prefer $8 - 6 = 2$ because it "translates" exactly. The words translate into symbols in the same order as you read the statement.

$$
\begin{array}{cccc}
8 & \text{exceeds} & 6 & \text{by } 2 \\
8 & - & 6 & = 2
\end{array}
$$

Here is one alternate solution.

Let

$$
\begin{aligned}
x &= \text{smaller part} \\
x + 7 &= \text{larger part}
\end{aligned}
$$

Then

$$
\begin{aligned}
x + (x + 7) &= 71 \\
2x + 7 &= 71 \\
2x &= 64 \\
x &= 32 \\
x + 7 &= 39
\end{aligned}
\quad \Bigg\} \text{ Answers}
$$

CONSECUTIVE INTEGER PROBLEMS

There are some special kinds of numbers often found in algebra problems. These are consecutive integers. They are generally qualified as positive. They may be

1. *Consecutive integers*

   Consecutive integers are, for example, 21, 22, 23. The difference between consecutive integers is *one*.

Example Represent three consecutive integers.

Let

$$\begin{aligned} x &= \text{1st consecutive integer} \\ x+1 &= \text{2nd consecutive integer} \\ x+2 &= \text{3rd consecutive integer} \end{aligned}$$

2. *Consecutive even integers*

Consecutive even integers might be 2, 4, 6. The difference between consecutive *even* integers is *two*. The first integer in the sequence has to be even, of course.

Example Three consecutive even integers.

Let

$$\begin{aligned} x &= \text{1st even integer} \\ x+2 &= \text{2nd even integer} \\ x+4 &= \text{3rd even integer} \end{aligned}$$

3. *Consecutive odd integers*

An example of consecutive *odd* integers is 5, 7, 9. Here we find the difference is also *two*. But the first integer is *odd*.

Example

Let

$$\begin{aligned} x &= \text{1st consecutive odd integer} \\ x+2 &= \text{2nd consecutive odd integer} \\ x+4 &= \text{3rd consecutive odd integer} \end{aligned}$$

Note that both the *even* integer problem and the *odd* integer problem are set up exactly the same. The difference is that x represents an *even* integer in one and an *odd* integer in the other.

EXAMPLE 4. Find three consecutive integers whose sum is 87.

Solution

Let

$$\begin{aligned} x &= \text{1st consecutive integer} \quad \text{(Smallest number)} \\ x+1 &= \text{2nd consecutive integer} \\ x+2 &= \text{3rd consecutive integer} \end{aligned}$$

Equation

$$x + (x+1) + (x+2) = 87$$
$$3x + 3 = 87$$
$$3x = 84$$
$$\left.\begin{aligned} x &= 28 \\ x+1 &= 29 \\ x+2 &= 30 \end{aligned}\right\} \quad \textit{Answers}$$

Check The numbers 28, 29, and 30 are consecutive integers. The sum of 28, 29, and 30 is 87.

EXAMPLE 5. Find three consecutive *even* integers such that the largest is three times the smallest.

Solution

Let
$$x = \text{1st consecutive even integer}$$
$$x + 2 = \text{2nd consecutive even integer}$$
$$x + 4 = \text{3rd consecutive even integer}$$

Equation statement: largest is three times the smallest.

$$x + 4 = 3x$$
$$-2x = -4$$
$$\left.\begin{aligned} x &= 2 \\ x+2 &= 4 \\ x+4 &= 6 \end{aligned}\right\} \quad \textit{Answers}$$

Check 2, 4, 6 are even, consecutive integers. The largest is three times the smallest. Thus, 6 = 2(3).

Always be sure that *all* the unknowns at the beginning of the problem are solved for at the end.

EXAMPLE 6. Four consecutive *odd* integers have a sum of 64. Find the integers.

Solution

Let
$$x = \text{1st consecutive odd integer}$$
$$x + 2 = \text{2nd consecutive odd integer}$$
$$x + 4 = \text{3rd consecutive odd integer}$$
$$x + 6 = \text{4th consecutive odd integer}$$

Equation

$$
\begin{aligned}
x + (x+2) + (x+4) + (x+6) &= 64 \\
4x + 12 &= 64 \\
4x &= 52 \\
\left.\begin{aligned} x &= 13 \\ x+2 &= 15 \\ x+4 &= 17 \\ x+6 &= 19 \end{aligned}\right\} &\quad \textit{Answers}
\end{aligned}
$$

Check The sum of 13, 15, 17, and 19 is 64.

Now it is time to try some problems by yourself. Remember:

1. Read the problem all the way through.
2. Find out what you are trying to solve for.
3. Set up the unknown with "Let x =" starting every problem. If possible, let x represent smallest unknown.
4. Read the problem again, a small step at a time, translating words into algebraic statements to set up the unknowns and to state the equation.

SUPPLEMENTARY NUMBER PROBLEMS

1. There is a number such that three times the number minus 6 is equal to 45. Find the number.
2. The sum of two numbers is 41. The larger number is one less than twice the smaller number. Find the numbers.
3. Separate 90 into two parts so that one part is four times the other part.
4. The sum of three consecutive integers is 54. Find the integers.
5. There are two numbers whose sum is 53. Three times the smaller number is equal to 19 more than the larger number. What are the numbers?
6. There are three consecutive odd integers. Three times the largest is seven times the smallest. What are the integers?

7. The sum of four consecutive even integers is 44. What are the numbers?

8. There are three consecutive integers. The sum of the first two is 35 more than the third. Find the integers.

9. A 25-foot long board is to be cut into two parts. The longer part is 1 foot more than twice the shorter part. How long is each part?

10. Mrs. Mahoney went shopping for some canned goods which were on sale. She bought three times as many cans of tomatoes as cans of peaches. The number of cans of tuna was twice the number of cans of peaches. If Mrs. Mahoney purchased a total of 24 cans, how many of each did she buy?

11. The first side of a triangle is 2 inches shorter than the second side. The third side is 5 inches longer than the second. If the perimeter of the triangle is 33 inches, how long is each side?

12. Kerrie and Shelly rode their bicycles four more than three times as many miles in the afternoon as in the morning on a trip to the lake. If the entire trip was 112 miles, how far did they ride in the morning and how far in the afternoon?

13. Mr. and Mrs. Patton and their daughter Carolyn own three cars. Carolyn drives 10 miles per week farther with her car than her father does with his. Mr. Patton drives twice as many miles per week as Mrs. Patton. If their total mileage per week is 160 miles, how many miles per week does each drive?

14. There were 10,483 people who attended a rock festival. If there were 811 more boys than girls, and 2481 fewer adults over 50 years of age than there were girls, how many of each group attended the festival?

15. In a 3-digit number, the hundreds digit is four more than the units digit and the tens digit is twice the hundreds digit. If the sum of the digits is 12, find the three digits. Write the number.

SOLUTIONS TO SUPPLEMENTARY NUMBER PROBLEMS

1. Let $x =$ number

Three times the number minus 6 is 45.

$$3x - 6 = 45$$
$$3x = 51$$
$$x = 17 \quad \textit{Answer}$$

2. Let $x =$ smaller number

$41 - x =$ larger number (Total minus x)

Larger number is twice smaller number less 1.

$$41 - x = 2x - 1$$
$$-3x = -42$$
$$\left.\begin{aligned} x &= 14 \\ 41 - x &= 27 \end{aligned}\right\} \quad \textit{Answers}$$

Alternate Solution

Let $x =$ smaller number

$2x - 1 =$ larger number (Larger number is one less than twice the smaller)

Sum of the two numbers is 41.

$$x + (2x - 1) = 41$$
$$3x = 42$$
$$\left.\begin{aligned} x &= 14 \\ 2x - 1 &= 27 \end{aligned}\right\} \quad \textit{Answers}$$

3. Let $x =$ smaller part

$90 - x =$ larger part (Total minus x)

The larger part is four times smaller part.

$$90 - x = 4x$$
$$-5x = -90$$
$$\left.\begin{aligned} x &= 18 \\ 90 - x &= 72 \end{aligned}\right\} \quad \textit{Answers}$$

Alternate Solution

Let

$$x = \text{smaller part}$$
$$4x = \text{larger part} \quad \text{(Larger part is four times the smaller)}$$

Sum of the two numbers is 90.

$$x + 4x = 90$$
$$5x = 90$$
$$\left.\begin{aligned} x &= 18 \\ 4x &= 72 \end{aligned}\right\} \textit{Answers}$$

4. Let

$$x = \text{1st consecutive integer}$$
$$x + 1 = \text{2nd consecutive integer}$$
$$x + 2 = \text{3rd consecutive integer}$$

Sum of three consecutive integers is 54.

$$x + (x+1) + (x+2) = 54$$
$$3x + 3 = 54$$
$$3x = 51$$
$$\left.\begin{aligned} x &= 17 \\ x+1 &= 18 \\ x+2 &= 19 \end{aligned}\right\} \textit{Answers}$$

5. Let

$$x = \text{smaller number}$$
$$53 - x = \text{larger number} \quad \text{(Total minus } x\text{)}$$

Three times smaller is larger plus 19.

$$3x = (53 - x) + 19$$
$$4x = 72$$
$$\left.\begin{aligned} x &= 18 \\ 53 - x &= 35 \end{aligned}\right\} \textit{Answers}$$

Alternate Solution

Let

$$x = \text{smaller number}$$
$$3x - 19 = \text{larger number} \quad \text{(Larger is 19 less than three times the smaller)}$$

Sum is 53.

$$x + 3x - 19 = 53$$
$$4x - 19 = 53$$
$$4x = 72$$

$$\left.\begin{aligned} x &= 18 \\ 3x - 19 &= 35 \end{aligned}\right\} \quad \textit{Answers}$$

6. Let

$$\left.\begin{aligned} x &= \text{1st consecutive odd integer} \\ x + 2 &= \text{2nd consecutive odd integer} \\ x + 4 &= \text{3rd consecutive odd integer} \end{aligned}\right\}$$

(Difference between consecutive odd integers is 2)

Three times the largest is seven times the smallest.

$$\begin{aligned} 3(x + 4) &= 7x \\ 3x + 12 &= 7x \\ -4x &= -12 \end{aligned}$$

$$\left.\begin{aligned} x &= 3 \\ x + 2 &= 5 \\ x + 4 &= 7 \end{aligned}\right\} \quad \textit{Answers}$$

7. Let

$$\left.\begin{aligned} x &= \text{1st consecutive } \textit{even} \text{ integer} \\ x + 2 &= \text{2nd consecutive } \textit{even} \text{ integer} \\ x + 4 &= \text{3rd consecutive } \textit{even} \text{ integer} \\ x + 6 &= \text{4th consecutive } \textit{even} \text{ integer} \end{aligned}\right\}$$

(Difference between consecutive even integers is 2)

Sum is 44.

$$\begin{aligned} x + (x + 2) + (x + 4) + (x + 6) &= 44 \\ 4x + 12 &= 44 \\ 4x &= 32 \\ x &= 8 \end{aligned}$$

$$\left.\begin{aligned} x + 2 &= 10 \\ x + 4 &= 12 \\ x + 6 &= 14 \end{aligned}\right\} \quad \textit{Answers}$$

8. Let

$$\left.\begin{aligned} x &= \text{1st consecutive integer} \\ x + 1 &= \text{2nd consecutive integer} \\ x + 2 &= \text{3rd consecutive integer} \end{aligned}\right\}$$

(Difference between consecutive integers is 1)

The sum of the first two is the third plus 35.

$$\begin{aligned} x + (x + 1) &= (x + 2) + 35 \\ 2x + 1 &= x + 37 \end{aligned}$$

$$\left.\begin{aligned} x &= 36 \\ x+1 &= 37 \\ x+2 &= 38 \end{aligned}\right\} \quad \textit{Answers}$$

9. Let

$$\begin{aligned} x &= \text{shorter part} \\ 25 - x &= \text{longer part} \end{aligned}$$

$$\begin{aligned} 25 - x &= 2x + 1 \\ -3x &= -24 \end{aligned}$$

$$\left.\begin{aligned} x &= 8 \\ 25 - x &= 17 \end{aligned}\right\} \quad \textit{Answers}$$

Alternate Solution

Let

$$\begin{aligned} x &= \text{shorter part} \\ 2x + 1 &= \text{longer part} \end{aligned}$$

$$\begin{aligned} x + 2x + 1 &= 25 \\ 3x &= 24 \end{aligned}$$

$$\left.\begin{aligned} x &= 8 \\ 2x + 1 &= 17 \end{aligned}\right\} \quad \textit{Answers}$$

Check The sum is 25 and $8 + 17 = 25$.

10. Let

$$\begin{aligned} x &= \text{number of cans of peaches} \quad \text{(Smallest number)} \\ 3x &= \text{number of cans of tomatoes} \\ 2x &= \text{number of cans of tuna} \end{aligned}$$

$$\begin{aligned} x + 3x + 2x &= 24 \\ 6x &= 24 \end{aligned}$$

$$\left.\begin{aligned} x &= 4 \\ 3x &= 12 \\ 2x &= 8 \end{aligned}\right\} \quad \textit{Answers}$$

Check The sum is 24 and $4 + 12 + 8 = 24$.

11. Let

$$\begin{aligned} x &= \text{length of 1st side in inches} \quad \text{(Shortest side)} \\ x + 2 &= \text{length of 2nd side in inches} \\ x + 2 + 5 &= \text{length of 3rd side in inches} \end{aligned}$$

$$\begin{aligned} x + (x + 2) + (x + 2 + 5) &= 33 \\ 3x + 9 &= 33 \\ 3x &= 24 \end{aligned}$$

$$\left.\begin{aligned} x &= 8 \\ x + 2 &= 10 \\ x + 2 + 5 &= 15 \end{aligned}\right\} \quad \textit{Answers}$$

Check The sum is 33 and $8 + 10 + 15 = 33$.

12. Let

x = number of miles in the morning (Smaller distance)

$3x + 4$ = number of miles in the afternoon

$$\begin{aligned} x + 3x + 4 &= 112 \\ 4x + 4 &= 112 \\ 4x &= 108 \end{aligned}$$

$$\left.\begin{aligned} x &= 27 \\ 3x + 4 &= 85 \end{aligned}\right\} \quad \textit{Answers}$$

Check The sum is 112 and $27 + 85 = 112$.

13. Let

x = number of miles Mrs. Patton drives (Smallest number)

$2x$ = number of miles Mr. Patton drives

$2x + 10$ = number of miles Carolyn drives

$$\begin{aligned} x + 2x + (2x + 10) &= 160 \\ 5x + 10 &= 160 \\ 5x &= 150 \end{aligned}$$

$$\left.\begin{aligned} x &= 30 \\ 2x &= 60 \\ 2x + 10 &= 70 \end{aligned}\right\} \quad \textit{Answers}$$

Check The sum is 160 and $30 + 60 + 70 = 160$.

14. Let

x = number of girls (Smallest group)

$x + 811$ = number of boys

$x - 2481$ = number of adults

$$\begin{aligned} x + x + 811 + x - 2481 &= 10{,}483 \\ 3x - 1670 &= 10{,}483 \\ 3x &= 12{,}153 \end{aligned}$$

$$\left.\begin{aligned} x &= 4051 \\ x + 811 &= 4862 \\ x - 2481 &= 1570 \end{aligned}\right\} \quad \textit{Answers}$$

Check The total number of people is 10,483.
$4051 + 4862 + 1570 = 10{,}483$.

15. Let

$$\begin{aligned} x &= \text{units digit} \quad \text{(Smallest number)} \\ x + 4 &= \text{hundreds digit} \\ 2(x + 4) &= \text{tens digit} \end{aligned}$$

$$\begin{aligned} x + (x + 4) + 2(x + 4) &= 12 \\ x + x + 4 + 2x + 8 &= 12 \\ 4x &= 0 \end{aligned}$$

$$\left.\begin{aligned} x &= 0 \\ x + 4 &= 4 \\ 2(x + 4) &= 8 \end{aligned}\right\} \quad \textit{Answers}$$

The number is 480.

Check The sum of the digits is 12 and $0 + 4 + 8 = 12$.

Chapter 2

Time, Rate, and Distance

Now that you know how to tackle a word problem and can work the number problems, let's try a new type of problem. Because most of you understand the time, rate, and distance relationship, let's learn a method for setting up time, rate, and distance problems.

First, a little review. If you travel for 2 hours at 50 mph to reach a destination, you know (I hope) that you would have traveled 100 miles. In short, time times rate equals distance or $t \times r = d$. It is convenient to have a diagram in time, rate, and distance problems. There are usually two moving objects. (Sometimes there is one moving object traveling at two different speeds.) Show in a small sketch the direction and distance of each movement. Then put the information in a simple diagram.

For example, one train leaves Chicago for Boston and at the same time another train leaves Boston for Chicago on the same track but traveling at a different speed. They travel until they meet. Your sketch would look like this:

Note that *times* are equal, but distances are unequal.

Or suppose two trains leave the same station at different times traveling in the same direction. One overtakes the other. The distances are equal at the point where one overtakes the other.

1st train

2nd train

Mr. Nemuth leaves New York for a drive to Long Island. Later he returns to New York.

New York → Long Island ← Distances are equal

The diagram for your time, rate, and distance information might look like this:

	Time	Rate	Distance
First object			
Second object			

You should fill in the appropriate information from each problem. Rate is in miles per hour and time is in *hours*. Distance will therefore be in miles. The basic relationship is *time times rate equals distance.*

EXAMPLE 1. A freight train starts from Los Angeles and heads for Chicago at 40 mph. Two hours later a passenger train leaves the same station for Chicago traveling 60 mph. How long before the passenger train overtakes the freight train?

Steps

1. Read the problem through, carefully.
2. The question at end of problem asks "how long?" (which means time) for passenger train. This question is your unknown.

Solution

First, draw sketch of movement:

Freight →
Los Angeles — Chicago
Passenger →
leaves 2 hr later

Notice that distances are equal.

Second, make a diagram to put in information:

	Time	Rate	Distance
Freight			
Passenger			

Now, read the problem again from the beginning for those individual steps. It says the freight train traveled 40 mph, so fill in the box for the rate of the freight:

	Time	Rate	Distance
Freight		40	
Passenger			

Continue reading: "Two hours later a passenger train... traveling 60 mph." Fill this figure in. You will always know either *both rates* or *both times* in the beginning problems.

	Time	Rate	Distance
Freight		40	
Passenger		60	

But so far you haven't any unknown. The question asked is, "How long before the passenger overtakes the freight?" As you know by now, you usually put the unknown down first. But when you use the time, rate, and distance table, it helps to put down known facts first because the problem tells you either both the times or both the rates.

Now put in x for the time for the passenger train. It's the unknown to be solved for. By the way, did you notice by your sketch that the *distances are equal* when one train overtakes the other? This fact is very important to remember.

Let x = time in hours for the passenger train

	Time	Rate	Distance
Freight		40	
Passenger	x	60	

Don't put anything in the distance box until all other information is filled in. Fill in *both rates* or *both times*. The other two will be filled in as unknowns. You will multiply time times rate to get distance and put this product in the distance box to represent distance. (Later you may have problems where *each* distance is given, but this type of problem results in fractions. For now, the distances will not be given.) Back to the problem. You have filled in both rates and one time. You must represent the time for the freight in terms of x because you know both *rates*, and therefore both times are unknown. The problem stated that the passenger train started 2 hours after the freight so the freight took 2 hours longer. You can represent the time for the freight by $x + 2$.

	Time	Rate	Distance
Freight	$x + 2$	40	
Passenger	x	60	

Now time times rate equals distance ($t \times r = d$) so multiply what you have in the time box times what you have in the rate box and put the result in the distance box:

	Time	Rate	Distance
Freight	$x + 2$	40	$40(x + 2)$
Passenger	x	60	$60x$

Every time, rate, and distance problem has some kind of relationship between the distances. This one had the distances equal. That is, the trains traveled the same distance because they started at the same place and traveled until one caught up with the other. This fact was not stated. You have to watch for the relationship. We have two distances in the table above. $40(x + 2)$ represents the distance for the freight train. $60x$ represents the distance for the passenger train. Set these two distances equal for your equation:

$$40(x+2) = 60x$$
$$40x + 80 = 60x$$
$$-20x = -80$$
$$x = 4 \text{ hours} \qquad \textit{Answer}$$

Check

$$40(4+2) = 60(4)$$
$$240 = 240$$

Some words of explanation. There was only *one* unknown asked for, but you had to use two to work the problem. Also, you could have used x for the freight's time and $x-2$ for the passenger train's time. Always be very careful that you have *answered the question.*

There are several types of time, rate, and distance problems. We have seen one object overtake another, starting from the same place at a different time. Here is another type:

EXAMPLE 2. A car leaves San Francisco for Los Angeles traveling an average of 60 mph. At the same time, another car leaves Los Angeles for San Francisco traveling 50 mph. If it is 440 miles between San Francisco and Los Angeles, how long before the two cars meet, assuming that each maintains its average speed?

Steps

1. Find out if you know both speeds or both times. In this problem we find that both speeds are given.
2. What is the unknown? The question is, "How long before the two cars meet?" So their time is unknown. But which shall we let x equal? *They leave at the same time* and *meet at the same time*, so the traveling times must be equal if neither one stops. Watch for the times being equal when two objects start at the same time and meet at the same time.
3. What fact is known about distance? The 440 miles is the *total* distance, so here is equation information. Remember, don't put distance information in the distance box unless you have to (for example, when the exact distance *each* object travels is given).

Solution

First, draw a sketch of movement:

Los Angeles ⟶ ⟵ San Francisco

Second, make a diagram and fill in all the information you have determined. And remember, time times rate equals distance in the table.

Let x = time in hours for two cars to meet

	Time	Rate	Distance
Car from Los Angeles	x	50	$50x$
Car from San Francisco	x	60	$60x$

You know that the total distance is 440 miles, so add the two distances in the diagram and set them equal to 440 miles for the equation.

$$60x + 50x = 440$$
$$110x = 440$$
$$x = 4 \text{ hours} \quad \textit{Answer}$$

Check

$$60(4) + 50(4) = 440$$
$$440 = 440$$

EXAMPLE 3. Two planes leave New York at 10 AM, one heading for Europe at 600 mph and one heading in the opposite direction at 150 mph. (So it isn't a jet!) At what time will they be 900 miles apart? How far has each traveled?

Steps

1. You know both *speeds.*
2. The times must be unknowns and they are equal (or the same) because the planes leave at the same time and travel until a certain given time (when they are 900 miles apart).
3. The total distance is 900 miles.

Warning! Do not let x = *clock* time, that is, the time of day. It stands for *traveling* time in hours! The rate is in miles per *hour,* so time must be in *hours* in all time, rate, distance problems.

Solution

Sketch:

New York

Slow plane | Fast plane

Let x = time in hours for planes to fly 900 miles apart

Remember, x equals the times for both the planes, since their times are the same in this problem.

Diagram:

	Time	Rate	Distance
Fast plane	x	600	$600x$
Slow plane	x	150	$150x$

Equation

$$600x + 150x = 900$$

$$750x = 900$$

$$x = \frac{900}{750}$$

$$x = 1\tfrac{1}{5} \text{ hours} \qquad \textit{Answer}$$

Is that what the problem asked for? Always check the question, especially on time, rate, and distance problems. The problem asked *what time* they would be 900 miles apart. Since the problem says they left at 10 AM, add $1\frac{1}{5}$ hours to that time. So the answer is 11:12 AM.

The problem also asked how far each would travel. Since we know that rate times time equals distance,

$$600x = \text{distance for fast plane}$$

and substituting the value of x,

$$600(1\tfrac{1}{5}) = 600(\tfrac{6}{5}) = 720 \text{ miles} \quad \textit{Answer}$$

$$150x = \text{distance for slow plane}$$

$$150(1\tfrac{1}{5}) = 150(\tfrac{6}{5}) = 180 \text{ miles} \quad \textit{Answer}$$

Check

$$720 + 180 = 900 \text{ miles}$$

EXAMPLE 4. Mr. Derbyshire makes a business trip from his house to Loganville in 2 hours. One hour later, he returns home in traffic at a rate 20 mph less than his rate going. If Mr. Derbyshire is gone a total of 6 hours, how fast did he travel on each leg of the trip?

Steps

1. This problem does not give us the rates, but it does give us the time both ways. It gives the time to Loganville as 2 hours. The total time is 6 hours, but 1 hour he did not travel. Deducting this hour leaves only 5 hours *total traveling time.* That means that if it took 2 hours to travel to Loganville it must have taken 3 hours to travel back.

2. The rates are both unknown. We can let x equal either one. Just be careful that the faster rate goes with the shorter time. If x equals rate going, then $x-20$ equals rate returning. (If x equals rate returning, then $x+20$ equals rate going. Either way is correct.)

Solution

Sketch:

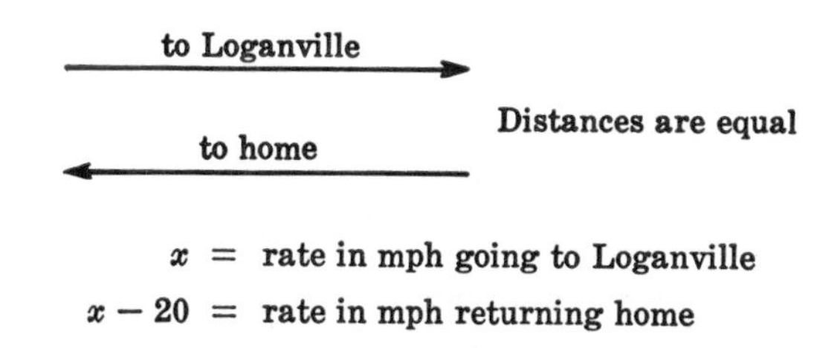

Let $\quad x = \text{rate in mph going to Loganville}$

$$x - 20 = \text{rate in mph returning home}$$

Diagram:

	Time	Rate	Distance
Toward Loganville	2	x	$2x$
Toward home	3	$x - 20$	$3(x - 20)$

Equation

The distance over equals distance back.

$$2x = 3(x - 20)$$
$$-x = -60$$
$$\left.\begin{aligned} x &= 60 \text{ mph, rate going} \\ x - 20 &= 40 \text{ mph, rate returning} \end{aligned}\right\} \quad \textit{Answers}$$

Check

$$2(60) = 3(60 - 20)$$
$$120 = 3(40)$$
$$120 = 120$$

EXAMPLE 5. Jake and Jerry went on a camping trip with their motorcycles. One day Jerry left camp on his motorcycle to go to the village. Ten minutes later Jake decided to go too. If Jerry was traveling 30 mph and Jake traveled 35 mph, how long before Jake caught up with Jerry?

Steps

1. We know both speeds.
2. We know that Jake traveled for 10 minutes *less* than Jerry. But minutes cannot be used in time, rate, and distance problems because rate is in miles per *hour*. So we must change 10 minutes into 10/60 or 1/6 of an hour.
3. They travel the same distance so the distances are equal.

Solution

Sketch:

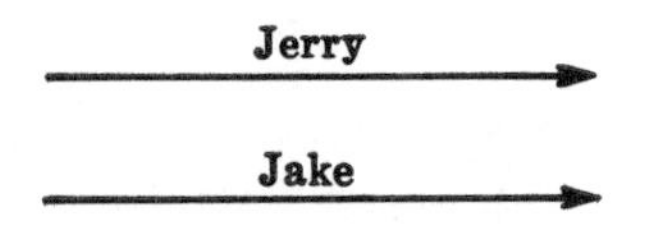

Let x = time in hours for Jake to catch Jerry

Diagram:

	Time	Rate	Distance
Jerry	$x + \frac{1}{6}$	30	$30(x + \frac{1}{6})$
Jake	x	35	$35x$

Equation

$$35x = 30\left(x + \frac{1}{6}\right)$$
$$35x = 30x + 5$$
$$5x = 5$$
$$x = 1 \text{ hour, Jake's time}$$ *Answer*

Alternate Solution

	Time	Rate	Distance
Jerry	x	30	$30x$
Jake	$x - \frac{1}{6}$	35	$35(x - \frac{1}{6})$

If you use the alternative solution, you should be good at fractions! Also, be sure to answer question and solve for $x - \frac{1}{6}$, Jake's time.

Note: Can you guess why this problem is here? You must be very careful to have *time always in hours* or fractions of an hour because rate is in miles per hour.

Basically, our time, rate, and distance problems look like these:

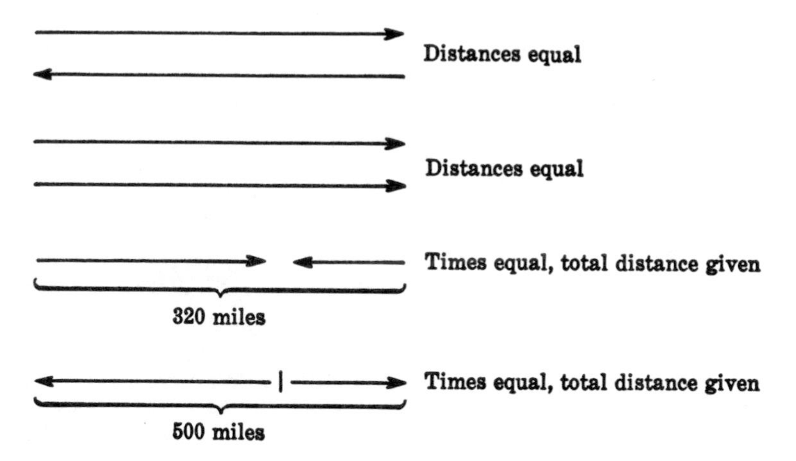

But some time, rate, and distance problems have tricky variations.

EXAMPLE 6. Two cars are headed for Las Vegas. One is 50 miles ahead of the other on the same road. The one in front is traveling 60 mph while the second car travels 70 mph. How long before the second car overtakes the first car?

Steps

1. The speeds are given.
2. The times are the unknowns and are equal.
3. The distances are not equal; this inequality is the tricky variation we mentioned above. But if we add 50 miles to the distance of one, it will equal the distance of the other.

Solution

Sketch:

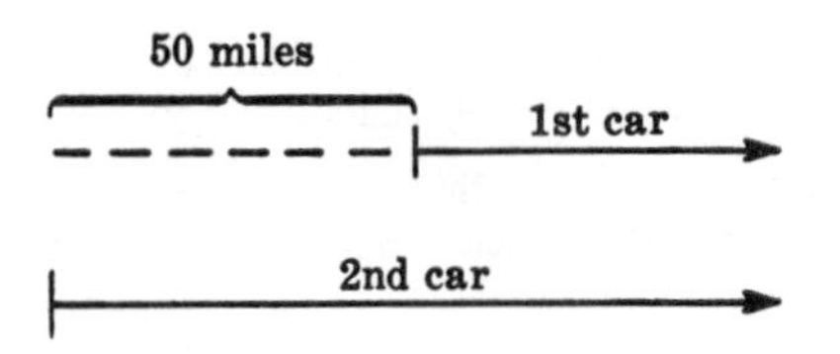

Let x = time in hours for second car to overtake first

Diagram:

	Time	Rate	Distance
First car	x	60	$60x$
Second car	x	70	$70x$

Equation

$$60x + 50 = 70x$$
$$-10x = -50$$
$$x = 5 \text{ hours} \qquad \textit{Answer}$$

If one car is 50 miles ahead of the other at the same time, it will take them the same length of time to reach the spot where they are together (or where one overtakes the other). So each takes x hours. The difficulty is that the two cars do not start from the same point, so their distances are not equal. The sketch shows that you have to add 50 miles to the distance the first one travels to make it equal to the distance the second one travels. Of course, the equation *could* be:

$$70x - 50 = 60x \qquad \text{or} \qquad 70x - 60x = 50$$

Any one of the three equations gives the same result, and any one is correct. If you can see the relationship better one way than another, always do the problem that way. Again it is like

$$6 + 2 = 8$$

$$8 - 2 = 6$$

$$8 - 6 = 2$$

All are correct statements of the relationship.

TIME, RATE, AND DISTANCE PROBLEMS INVOLVING MOVING AIR (WIND) OR MOVING WATER (CURRENT)

Some more difficult problems have planes flying in a wind or boats traveling in moving water. The only problems of this type which *we* can solve are those where the objects move directly *with* or *against* the wind or water. The plane must have a direct headwind or tailwind, and the boat must be going upstream or downstream. In this type of problem the plane's speed in still air would be increased by a tailwind or decreased by a headwind to determine how fast it actually covers the ground. For example, a plane flies 200 mph in still air. This is called airspeed. If there is a 20 mph headwind blowing, it would decrease the speed over the ground by 20 mph so the *groundspeed* of the plane would be 200 – 20 or 180 mph. The *groundspeed* is the *rate* in time, rate, and distance problems. A headwind *reduces* the speed of the plane by the velocity of the wind. A tailwind *increases* the speed of the plane over the ground by the velocity of the wind. A plane with an airspeed (speed in still air) of 400 mph with a 30 mph tailwind actually travels over the ground (groundspeed) at 430 mph. A current affects a boat in the same way.

EXAMPLE 7. A plane takes 5 hours to fly from Los Angeles to Honolulu and $4\frac{1}{11}$ hours to return from Honolulu to Los Angeles. If the wind velocity is 50 mph from the west on both trips, what is the airspeed of the plane? (Airspeed is speed of plane in still air.)

Steps

1. The two times are given.
2. You are asked to find the speed of the plane in *still* air (airspeed).
3. Going to Honolulu you have a *headwind* so *subtract* the velocity of the wind. Returning to Los Angeles, you have a *tailwind* so *add* the wind velocity to the airspeed.
4. The distances are equal.

Solution

Sketch: (use map directions)

wind → 50 mph

Honolulu ← Headwind — Los Angeles

Honolulu — Tailwind → Los Angeles

Let x = speed of plane in still air (airspeed)

$x - 50$ = speed of plane over the ground on trip from Los Angeles to Honolulu

$x + 50$ = speed of plane over the ground on trip from Honolulu to Los Angeles

Groundspeed determines how long it takes the plane to travel from one place to another.

Diagram:

	Time	Rate	Distance
To Honolulu against wind	5	$x - 50$	$5(x - 50)$
To Los Angeles with wind	$4\frac{1}{11}$	$x + 50$	$4\frac{1}{11}(x + 50)$

Equation

The distances are equal.

$$5(x - 50) = 4\tfrac{1}{11}(x + 50)$$

$$5(x - 50) = \tfrac{45}{11}(x + 50)$$

$$\frac{5x - 250}{1} = \frac{45x + 2250}{11}$$

Multiply both sides by 11 to clear fractions.

$$55x - 2750 = 45x + 2250$$

$$10x = 5000$$

$$x = 500 \text{ mph, airspeed} \qquad \textit{Answer}$$

Check

$$5(500 - 50) = \tfrac{45}{11}(500 + 50)$$

$$2250 = \frac{24{,}750}{11}$$

$$2250 = 2250$$

The plane will fly at the same *airspeed* regardless of the wind velocity. The speed at which it actually covers the ground is the airspeed plus or minus wind velocity, assuming a direct tailwind or headwind.

EXAMPLE 8. In his motor boat, a man can go downstream in 1 hour less time than he can go upstream the same distance. If the current is 5 mph, how fast can he travel in still water if it takes him 2 hours to travel upstream the given distance?

Steps

1. The times are 2 hours upstream and 1 hour downstream.
2. The rates are unknown. If you let x equal his rate of traveling in still water, his rate upstream will be $x - 5$ and his rate downstream will be $x + 5$. You subtract or add rate of current.
3. The distance up equals the distance down.

Solution

Sketch:

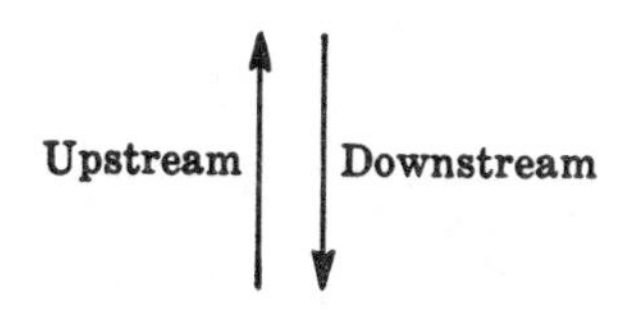

Let $\quad x =$ rate of boat in mph in still water

$x - 5 =$ rate of boat in mph upstream

$x + 5 =$ rate of boat in mph downstream

Diagram:

	Time	Rate	Distance
Up	2	$x - 5$	$2(x - 5)$
Down	1	$x + 5$	$1(x + 5)$

Equation

$$2(x - 5) = 1(x + 5)$$
$$2x - 10 = x + 5$$
$$x = 15 \text{ mph, rate in still water}$$ *Answer*

Check

$$2(15 - 5) = 1(15 + 5)$$
$$20 = 20$$

Notes to remember about time, rate, and distance problems:

1. Don't let x equal distance unless you have to. (This will sometimes be necessary when you get to fraction problems when *each* distance is given.)
2. You will always be given either *both rates* or *both times* (or later, both distances).
3. *Time* has to be in *hours* or fractions of an hour. You can never use minutes with mph.
4. Check sketch to see if distances are equal for equation information.
5. If two objects start at the same time, their traveling times are generally equal.

SUPPLEMENTARY TIME, RATE, AND DISTANCE PROBLEMS

1. Sarah leaves Seattle for New York in her car, averaging 80 mph across open country. One hour later a plane leaves Seattle for New York following the same route and flying 400 mph. How long before the plane overtakes the car?
2. A train averaging 50 mph leaves San Francisco at 1 PM for Los Angeles 440 miles away. At the same time a second train leaves Los Angeles headed for San Francisco on the same track and traveling at an average rate of 60 mph. At what time does the accident occur?

3. Mr. Kasberg rides his bike at 6 mph to the bus station. He then rides the bus to work, averaging 30 mph. If he spends 20 minutes less time on the bus than on the bike, and the distance from his house to work is 26 miles, what is the distance from his house to the bus station?

4. Two planes leave Dingle City at 1 PM. Plane A heads east at 450 mph and Plane B heads due west at 600 mph. How long will it be before the planes are 2100 miles apart?

5. Doomtown is 200 miles due west of Sagebrush and Joshua is due west of Doomtown. At 9 AM Mr. Archer leaves Sagebrush for Joshua. At 1 PM Mr. Sassoon leaves Doomtown for Joshua. If Mr. Sassoon travels at an average speed 20 mph faster than Mr. Archer and they each reach Joshua at 4 PM, how fast is each traveling?

6. Felicia left Rome at 8 AM and drove her Ferrari at 80 mph from Rome to Sorrento. She then took the boat to Capri for the day, returning to Sorrento 5 hours later. On the return trip from Sorrento to Rome she averaged 60 mph and arrived in Rome at 8 PM. How far is it from Rome to Sorrento?

7. Robin flies to San Francisco from Santa Barbara in 3 hours. She flies back in 2 hours. If the wind is blowing from the north at a velocity of 40 mph going, but changed to 20 mph from the north returning, what was the airspeed of the plane (its speed in still air)?

8. Timothy leaves home for Skedunk 400 miles away. After 2 hours, he has to reduce his speed by 20 mph due to rain. If he takes 1 hour for lunch and gas, and reaches Skedunk 9 hours after he left home, what was his initial speed?

9. A highway patrolman spots a speeding car. He clocks it at 70 mph and takes after it 0.5 mile behind. If the patrolman travels at an average rate of 90 mph, how long before he overtakes the car?

10. Boy Scouts hiking in the mountains divide into two groups to hike around Lake Sahara. They leave the same place at 10 AM, and one group of younger boys hikes east around the lake and the other group hikes west. If the younger boys hike at a rate of 3 mph and the older boys hike at a rate of 5 mph, how long before they meet on the other side of the lake if the trail around the lake is 8 miles long? At what time do they meet?

11. A boat has a speed of 15 mph in still water. It travels downstream from Greentown to Glenavon in $\frac{2}{5}$ of an hour. It then goes back upstream from Glenavon to Cambria, which is 2 miles downstream from Greentown, in $\frac{3}{5}$ of an hour. Find the rate of the current.

12. Carlos can run the mile in 6 minutes. Kevin can run the mile in 8 minutes. If Kevin goes out to practice and starts 1 minute before

Carlos starts on his practice run, will Carlos catch up with Kevin before he reaches the mile marker? (Assume straight track and average speed.)

13. Jeanette rides her bike to the bus station where she barely makes it in time to catch the bus to work. She spends half an hour on her bike and two-thirds of an hour on the bus. If the bus travels 39 mph faster than she travels on her bike, and the total distance from home to work is 40 miles, find the rate of the bike and the rate of the bus.

SOLUTIONS TO SUPPLEMENTARY TIME, RATE, AND DISTANCE PROBLEMS

1. Sketch:

Let

x = time in hours for plane to travel same distance as car

Diagram:

	Time	Rate	Distance
Car	$x+1$	80	$80(x+1)$
Plane	x	400	$400x$

Plane leaves 1 hour later, so car travels 1 hour longer

Equation

$$80(x+1) = 400x$$
$$80x + 80 = 400x$$
$$-320x = -80$$
$$x = \frac{80}{320}$$
$$x = \frac{1}{4} \text{ hour} \qquad \textit{Answer}$$

Check

$$80\left(1+\frac{1}{4}\right) = 400\left(\frac{1}{4}\right)$$
$$80\left(\frac{5}{4}\right) = 400\left(\frac{1}{4}\right)$$
$$100 = 100$$

Alternate Solution

Let

x = time in hours car travels before plane overtakes it

Diagram:

	Time	Rate	Distance
Car	x	80	$80x$
Plane	$x-1$	400	$400(x-1)$

Plane takes 1 hour less time

$$80x = 400(x+1)$$
$$80x = 400x - 400$$
$$-320x = -400$$
$$\left.\begin{aligned} x &= 1\tfrac{1}{4} \text{ hours} \\ x-1 &= \frac{1}{4} \text{ hour} \end{aligned}\right\} \quad \textit{Answers}$$

2. Sketch:

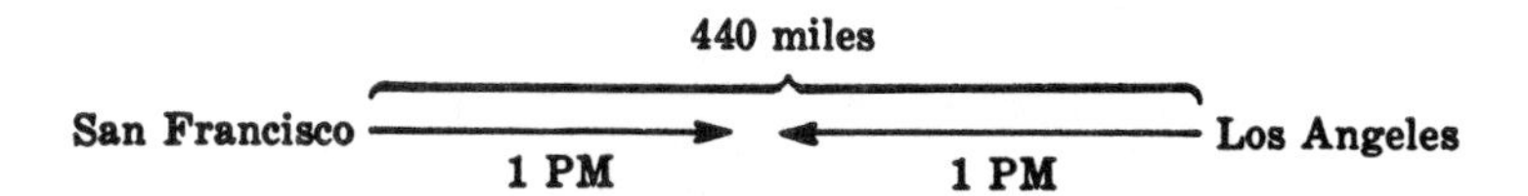

Let x = time in hours for trains to meet

Diagram:

	Time	Rate	Distance
First train	x	50	$50x$
Second train	x	60	$60x$

Trains leave at same time and arrive at meeting point at same time, so times are the same.

Equation

$$50x + 60x = 440$$

$$x = 4 \text{ hours}$$

Since the trains left at 1 PM, they meet 4 hours later at 5 PM. *Answer*

Check

$$50(4) + 60(4) = 440$$
$$200 + 240 = 440$$

3. Sketch:

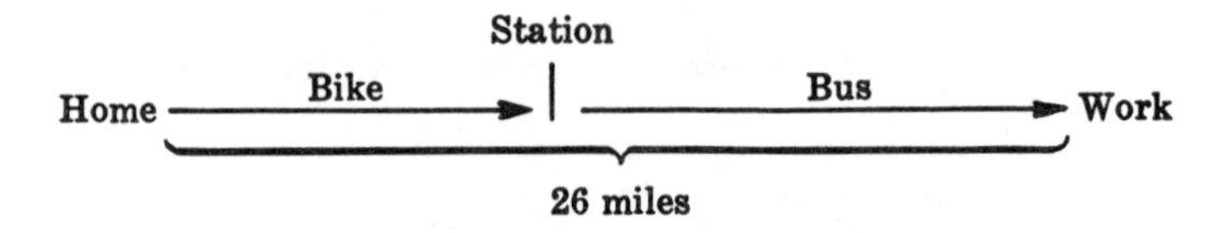

Let x = time in hours from home to bus station

Diagram:

	Time	Rate	Distance
Bike	x	6	$6x$
Bus	$x - \frac{1}{3}$	30	$30(x - \frac{1}{3})$

(20 minutes is $\frac{1}{3}$ hour)

Equation

$$6x + 30\left(x - \frac{1}{3}\right) = 26$$

$$6x + 30x - 10 = 26$$

$$x = 1 \text{ hour}$$

Since the problem asked the distance from house to bus station, we multiply 6 times x,

$$6x = 6 \text{ miles} \quad \textit{Answer}$$

Check

$$6(1) + 30\left(1 - \frac{1}{3}\right) = 26$$

$$6 + 20 = 26$$

4. Sketch:

2100 miles

Plane B (1 PM) Plane A

Dingle City

Let

x = time in hours each plane travels

Diagram:

	Time	Rate	Distance
Plane A	x	450	$450x$
Plane B	x	600	$600x$

Equation

$$450x + 600x = 2100$$
$$1050x = 2100$$
$$x = 2 \text{ hours} \qquad \textit{Answer}$$

5. Sketch:

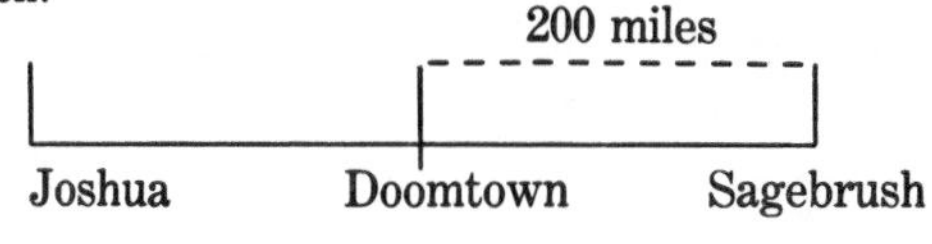

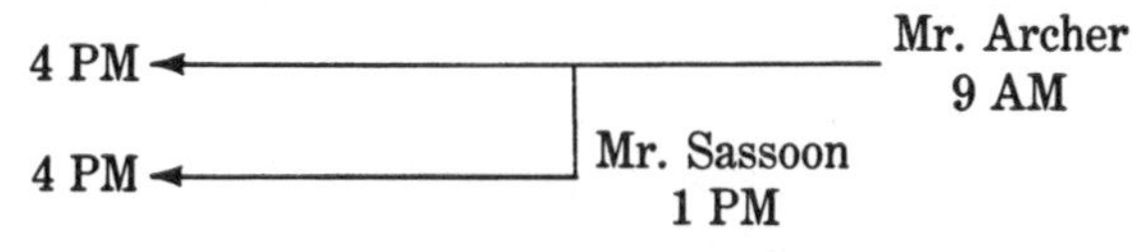

Let x = Mr. Archer's speed in mph
$x + 20$ = Mr. Sassoon's speed in mph

Diagram:

	Time	Rate	Distance
Archer	7	x	$7x$
Sassoon	3	$x + 20$	$3(x + 20)$

Equation

Mr. Archer travels 200 miles farther than Mr. Sassoon.

$$7x = 3(x + 20) + 200$$
$$7x = 3x + 60 + 200$$
$$4x = 260$$
$$x = 65 \text{ mph, Mr. Archer's speed}$$
$$x + 20 = 85 \text{ mph, Mr. Sassoon's speed}$$
Answers

Check

$$7(65) = 3(35) + 200$$
$$455 = 255 + 200$$
$$455 = 455$$

6. Sketch:

8 AM →

Rome Sorrento (5 hr time out)

8 PM ←

Let x = time in hours from Rome to Sorrento

	Time	Rate	Distance
Go	x	80	$80x$
Return	$7 - x$	60	$60(7 - x)$

Traveling time is total time (12 hours) less 5 hours out for Capri, or *7 hours*. Time on return trip is total traveling time minus time on trip to Sorrento.

Equation

$$80x = 60(7 - x)$$

$$80x = 420 - 60x$$

$$x = 3 \text{ hours}$$

Since the problem asked the distance from Rome to Sorrento, we multiply 80 times x,

$$80x = 80(3) = 240 \text{ miles} \quad \textit{Answer}$$

Check

$$80(3) = 60(7 - 3)$$

$$240 = 60(4)$$

$$240 = 240$$

7. Sketch:

wind going ↓ 40 mph

wind returning ↓ 20 mph

Let

x = speed of plane in still air (airspeed)

40 = velocity of wind from north (headwind) going

$x - 40$ = groundspeed of plane going to San Francisco against wind

20 = velocity of wind from north (tailwind) returning

$x + 20$ = groundspeed returning to Santa Barbara

Diagram:

	Time	Rate	Distance
SB to SF	3	$x - 40$	$3(x - 40)$
SF to SB	2	$x + 20$	$2(x + 20)$

Equation

Distance going equals distance returning.

$$3(x - 40) = 2(x + 20)$$
$$3x - 120 = 2x + 40$$
$$x = 160 \text{ mph, airspeed} \qquad \textit{Answer}$$

Check

$$3(160 - 40) = 2(160 + 20)$$
$$360 = 360$$

8. Sketch:

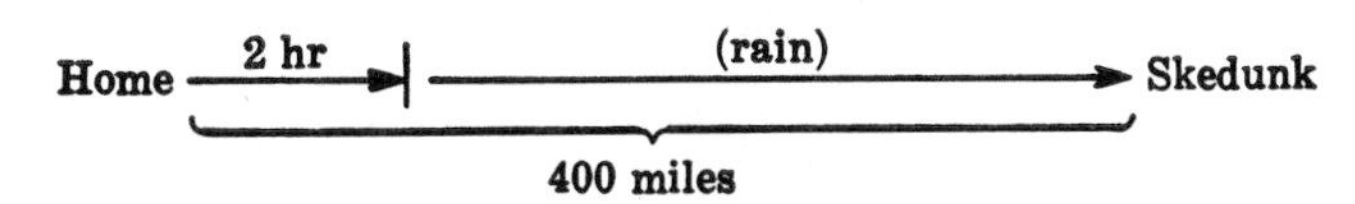

Let x = Timothy's original speed in mph

Diagram:

	Time	Rate	Distance
First	2	x	$2x$
Second	6	$x-20$	$6(x-20)$

Total time of 9 hours, less 2 hours first part, less 1 hour stopped, leaves 6 hours traveling time in the rain.

Equation

$$2x + 6(x-20) = 400$$
$$8x - 120 = 400$$
$$8x = 520$$
$$x = 65 \text{ mph} \qquad \textit{Answer}$$

Check

$$2(65) + 6(65-20) = 400$$
$$130 + 270 = 400$$
$$400 = 400$$

9. Sketch:

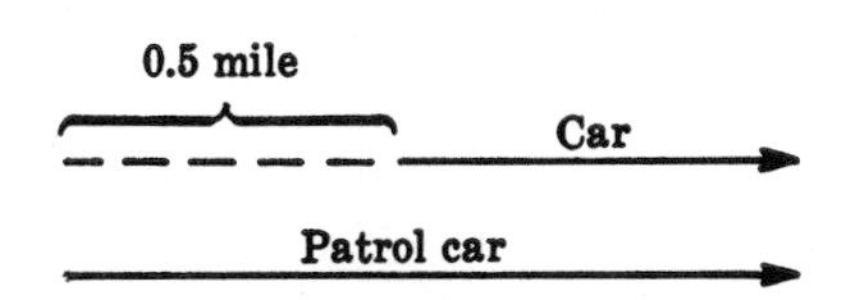

Let

x = time in hours for patrol car to overtake car

Diagram:

	Time	Rate	Distance
Car	x	70	$70x$
Patrol car	x	90	$90x$

Equation

The patrol car travels 0.5 mile farther than other car.

$$90x = 70x + 0.5$$

$$20x = 0.5$$

Multiply by 10 to clear decimal.

$$200x = 5$$

$$x = \frac{1}{40} \text{ of an hour} \qquad \textit{Answer}$$

If you wish, change the $\frac{1}{40}$ hour to minutes, $\frac{1}{40} \times 60 = 1\frac{1}{2}$ minutes.

Check

$$90\left(\frac{1}{40}\right) = 70\left(\frac{1}{40}\right) + \frac{1}{2}$$

$$\frac{90}{40} = \frac{70}{40} + \frac{20}{40}$$

$$\frac{90}{40} = \frac{90}{40}$$

10. Sketch:

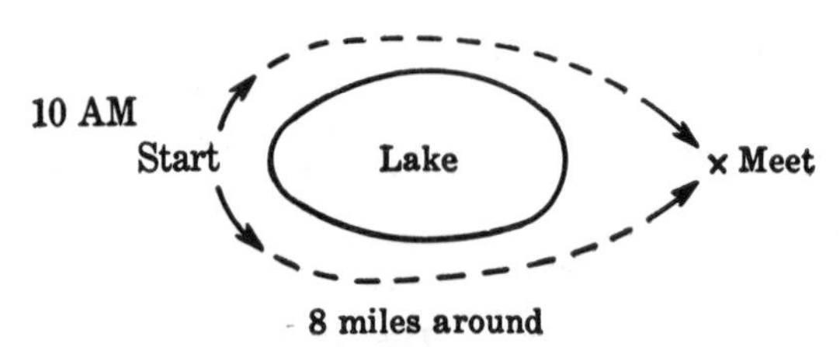

Let x = time in hours boys take to meet

Diagram:

	Time	Rate	Distance
Young Scouts	x	3	$3x$
Older Scouts	x	5	$5x$

The distance the younger Scouts traveled plus the distance the older Scouts traveled equals the total distance around lake.

Equation

$$3x + 5x = 8$$
$$8x = 8$$
$$x = 1 \text{ hour} \qquad \textit{Answer}$$

Since the problem also asks what time they meet,

$$10 \text{ AM} + 1 \text{ hour} = 11 \text{ AM} \qquad \textit{Answer}$$

Check

$$3x + 5x = 8$$
$$3(1) + 5(1) = 8$$
$$3 + 5 = 8$$
$$8 = 8$$

11. Sketch:

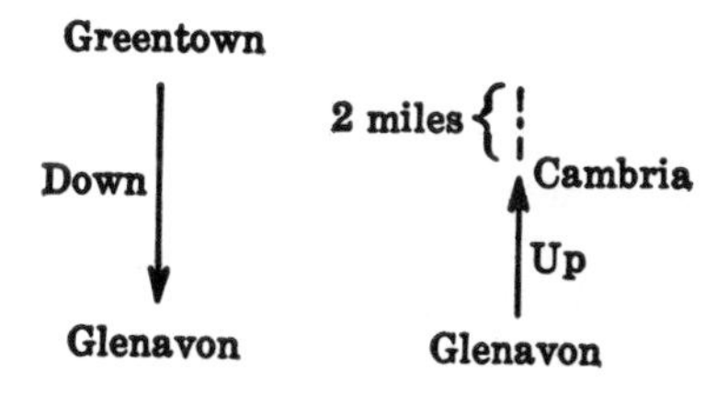

Let x = rate of current in mph

$15 + x$ = rate of boat downstream in mph

$15 - x$ = rate of boat upstream in mph

Diagram:

	Time	Rate	Distance
Downstream	$\frac{2}{5}$	$x + 15$	$\frac{2}{5}(15 + x)$
Upstream	$\frac{3}{5}$	$x - 15$	$\frac{3}{5}(15 - x)$

Equation

Distance from Greentown to Glenavon is equal to distance from Glenavon to Cambria plus 2 miles.

$$\frac{2}{5}(15 + x) = \frac{3}{5}(15 - x) + 2$$

Remove parentheses.

$$\frac{30 + 2x}{5} = \frac{45 - 3x}{5} + 2$$

Multiply each term by LCD, 5, to clear fractions.

$$30 + 2x = 45 - 3x + 5(2)$$
$$5x = 25$$
$$x = 5 \text{ mph} \qquad \textit{Answer}$$

Check

$$\frac{2}{5}(15 + x) = \frac{3}{5}(15 - x) + 2$$
$$\frac{2}{5}(15 + 5) = \frac{3}{5}(15 - 5) + 2$$
$$\frac{2}{5}(20) = \frac{3}{5}(10) + 2$$
$$8 = 6 + 2$$
$$8 = 8$$

12. Sketch:

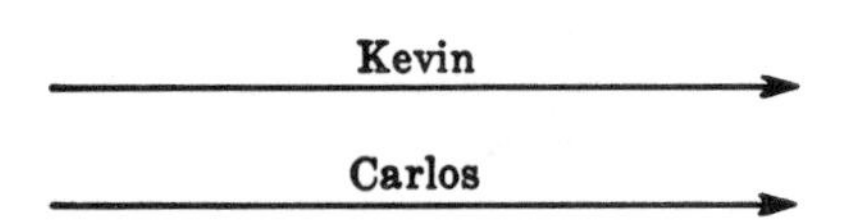

Convert minutes into fractions of an hour first.

$$6 \text{ minutes} = \frac{6}{60} \text{ or } \frac{1}{10} \text{ hours to run 1 mile}$$
$$8 \text{ minutes} = \frac{8}{60} \text{ or } \frac{2}{15} \text{ hours to run 1 mile}$$

Convert the time to run 1 mile into mph by multiplying it by a quantity which turns the time into 1 hour. This quantity is then equivalent to the miles run in 1 hour, or mph.

$$\frac{1}{10} \text{ hours} \times 10 = 1 \text{ hour} \qquad 10 \times 1 \text{ mile} = 10 \text{ mph}$$
$$\frac{2}{15} \text{ hours} \times \frac{15}{2} = 1 \text{ hour} \qquad \frac{15}{2} \times 1 \text{ mile} = 7\tfrac{1}{2} \text{ mph}$$

Thus Kevin's rate was $7\frac{1}{2}$ mph and Carlos' rate was 10 mph.

Let x = time in hours for Carlos to catch Kevin

Diagram:

	Time	Rate	Distance
Carlos	x	10	$10x$
Kevin	$x + \frac{1}{60}$	$\frac{15}{2}$	$\frac{15}{2}(x + \frac{1}{60})$

(Kevin takes one minute more time than Carlos. One minute is $\frac{1}{60}$ of an hour.)

Distances are equal in this problem, and we can use this equality for our equation.

$$10x = \frac{15}{2}\left(x + \frac{1}{60}\right)$$

$$10x = \frac{15x}{2} + \frac{1}{8}$$

$$80x = 60x + 1$$

$$20x = 1$$

$$x = \frac{1}{20} \text{ of an hour}$$

Therefore it takes Carlos $\frac{1}{20}$ hour to catch Kevin. Carlos' distance is $10x$. Substituting the value of x to get distance in miles,

$$10x = 10 \times \frac{1}{20} = \frac{1}{2}$$

Therefore $\frac{1}{2}$ mile from the start Carlos will catch up. Thus, Carlos will catch up with Kevin before he reaches the mile marker.

Answer

Check

$$10x = \frac{15}{2}\left(x + \frac{1}{60}\right)$$

$$10\left(\frac{1}{20}\right) = \frac{15}{2}\left(\frac{1}{20} + \frac{1}{60}\right)$$

$$\frac{1}{2} = \frac{15}{2}\left(\frac{3}{60} + \frac{1}{60}\right)$$

$$\frac{1}{2} = \frac{15}{2}\left(\frac{4}{60}\right)$$

$$\frac{1}{2} = \frac{1}{2}$$

13. Sketch:

Let x = rate of bicycle in mph

$x + 39$ = rate of bus in mph

Diagram:

	Time	Rate	Distance
Bike	$\frac{1}{2}$	x	$\frac{1}{2}x$
Bus	$\frac{2}{3}$	$x + 39$	$\frac{2}{3}(x + 39)$

(Write $\frac{1}{2}x$ as $\frac{x}{2}$ in your equation)

Equation

Distance on bike *plus* distance on bus equals total distance.

$$\frac{x}{2} + \frac{2(x + 39)}{3} = 40$$

Remove parentheses.

$$\frac{x}{2} + \frac{2x + 78}{3} = 40$$

Multiply by LCD, 6, to clear fractions.

$$6\left(\frac{x}{2}\right) + 6\left(\frac{2x + 78}{3}\right) = 6(40)$$

$$3x + 4x + 156 = 240$$

$$7x = 84$$

$$\left.\begin{aligned} x &= 12 \text{ mph} \\ x + 39 &= 51 \text{ mph} \end{aligned}\right\} \quad \textit{Answers}$$

Check

$$\frac{12}{2} + \frac{2(12 + 39)}{3} = 40$$

$$6 + \frac{102}{3} = 40$$

$$6 + 34 = 40$$

$$40 = 40$$

Chapter 3

Mixtures

Now let's tackle a new type of problem—mixture problems. They sound complicated, but we can set them up very methodically. There are two types of mixture problems, those dealing with percent and those involving price. The mechanics of the problems are exactly the same, even though they sound very different. Let us try the percent problems first. There are two important facts in mixture problems involving percents: the *percent* in each mixture and the *amount*. You will *always* be given all *three* percents and *one* amount in the beginning problems. You will be taking two mixtures and dumping them into one pot to make a third mixture. You will know the percent in each of the two mixtures and also the percent in the total mixture. You will also be given the amount in one of them.

The equation for a mixture problem involving percent when all the percents are given is based on the following fact: Amount in the mixture times the percent of pure stuff equals the *amount* of pure stuff (i.e., pure boric acid, pure silver, pure alcohol, etc.).

Example A mixture containing 20 quarts which is 15% alcohol would contain

$$20(0.15) = 3 \text{ quarts } \textit{pure} \text{ alcohol}$$

If you mix two solutions, the amount of *pure* stuff in the first plus the amount of *pure* stuff in the second will equal the amount of *pure* stuff in the *total* mixture. (The *percent* of pure stuff in the total mixture is *not* equal to the sum of the other two percents.)

Remember these facts:

1. Amount in first mixture plus amount in second mixture equals amount in total mixture. We will use this fact to set up unknowns.
2. Amount of pure stuff in first mixture plus amount of pure stuff in second mixture equals amount of pure stuff in total mixture. This is the *equation* information. It will look like this:

   $$(2\%)(12 \text{ quarts}) + (5\%)(x \text{ quarts}) = (3\%)[(12+x) \text{ quarts}]$$

   or $$0.02(12) + 0.05(x) = 0.03(12+x)$$

   Percent times amount plus percent times amount equals percent times amount.

Now let's try a problem.

EXAMPLE 1. A mixture containing 6% boric acid is to be mixed with 2 quarts of a mixture which is 15% acid in order to obtain a solution which is 12% acid. How much of the 6% solution must be used?

Steps

1. Draw three diagrams to represent the three solutions. They can be squares, circles, etc. We shall use circles.

1st mixture + 2nd mixture = Total mixture

2. Find out what the percents are in each original mixture and in the combined mixture and put them in the circles:

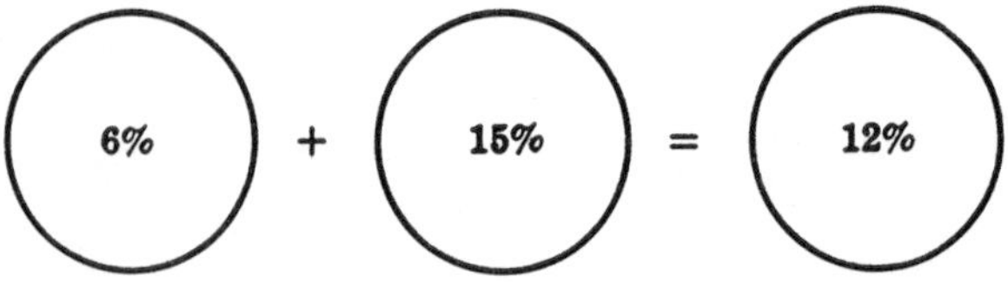

6% + 15% = 12%

3. Find the *one given amount* and put it in the proper circle:

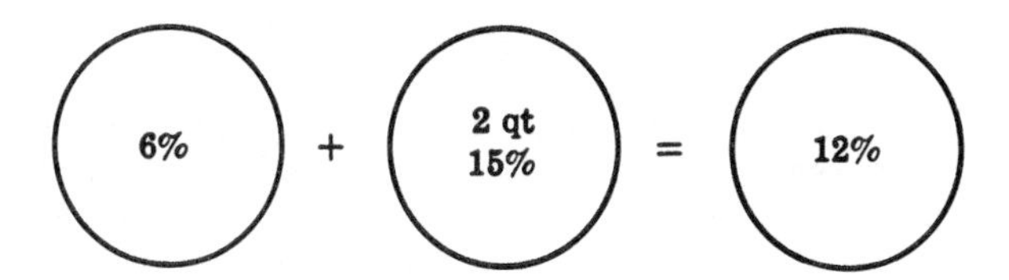

4. Represent the other two amounts using unknowns. The question asks, "How much of the 6% solution must be used?" so that is what x represents. The total solution will have x quarts plus 2 quarts (or the total of the other two solutions).

   Let x = quarts of 6% solution

x qt 6% + 2 qt 15% = $(x + 2)$ qt 12%

You are now ready to solve the problem. If you multiply the amount of solution by the percent of acid in the solution you will find the amount of *pure* boric acid in each solution. The amount of pure acid in the final solution is *equal* to the amounts of pure acid in the two original solutions added. This equality gives you the equation information. Keep *all amounts on top* to avoid confusion. Show products of amount times percent in lower part of diagram. These products represent the amount of pure acid in each mixture.

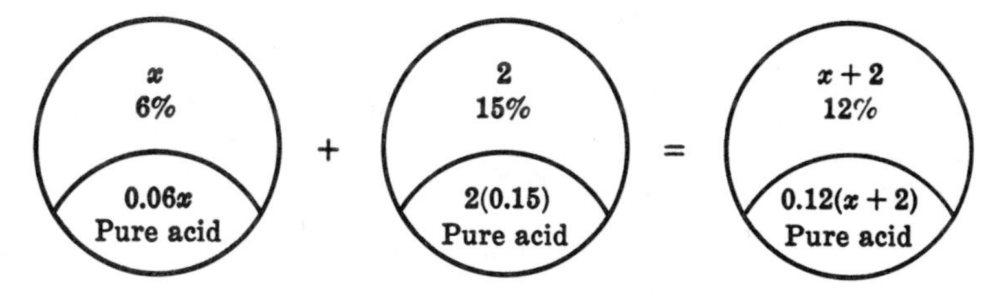

Equation $0.06x + 2(0.15) = 0.12(x + 2)$

It is safer to *eliminate parentheses first* and second eliminate decimals before proceeding to solve for x.

$$0.06x + 0.30 = 0.12x + 0.24$$

To clear decimals, multiply by a power of ten which will make all numbers whole numbers. Here we multiply by 100 which moves all decimal points *two* places to the right (same as number of zeros in 100).

$$6x + 30 = 12x + 24$$
$$-6x = -6$$
$$x = 1 \quad \textit{Answer}$$

Check

$$0.06x + 0.30 = 0.12x + 0.24$$
$$0.06(1) + 0.30 = 0.12(1) + 0.24$$
$$0.36 = 0.36$$

EXAMPLE 2. A farmer wants to mix milk containing 3% butterfat with cream containing 30% butterfat to obtain 900 gallons of milk which is 8% butterfat. How much of each must he use?

Steps

1. Draw diagram and fill in the known percents and the one known amount. Keep *amount* on *top* in each diagram.

3% + 30% = 900 gal 8%

2. Determine what x stands for. Here it will represent amount of one mixture. There are two mixtures with unknown amounts and it doesn't make any difference which amount is x.

Let x = gallons in 3% mixture

$900 - x$ = gallons in 30% mixture

x gal, 3% + 30% = 900, 8%

3. Represent the amount in the other mixture using x and the amount given in the problem. Remember that the amounts in the two original mixtures add up to the amount in the total mixture. The second unknown is $900 - x$. (Another example of the use of $total - x$ to represent second unknown.)

x, 3% + $(900 - x)$ gal, 30% = 900, 8%

4. Multiply each percent by the amount in its mixture and set these products equal:

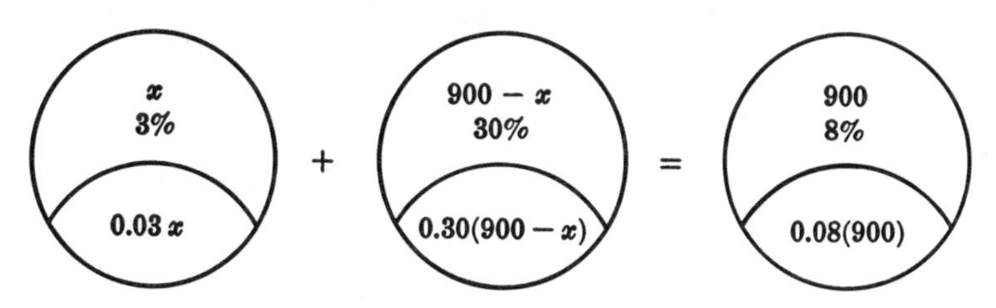

Equation

$$0.03x + 0.30(900 - x) = 0.08(900) \quad \text{Pure butterfat}$$

Clear parentheses.

$$0.03x + 270 - 0.30x = 72$$

Multiply by 100.

$$3x + 27{,}000 - 30x = 7200$$

$$-27x = -19{,}800$$

$$\left.\begin{aligned} x &= 733\tfrac{1}{3} \\ 900 - x &= 166\tfrac{2}{3} \end{aligned}\right\} \quad \textit{Answers}$$

EXAMPLE 3. Now let's look at a mixture problem involving prices instead of percents. For example, a merchant wishes to mix two grades of coffee, one of which sells for \$0.80 per pound and one of which sells for \$1.20 per pound. He wants to sell the mixture for \$1.10 per pound. If he has 25 pounds of the 80¢ coffee, how much of the \$1.20 per pound coffee must he add so that the value of the final mixture is equal to the total value of the other two?

This problem sounds different from the first type, but like the percent problem it has certain basic facts. There are two quantities mixed. There are *three prices given* instead of 3 percents. *One* amount is given. The problem can be illustrated by the same type of diagram.

Steps

1. Draw three circles (or squares or rectangles) to represent the three types of coffee and label the price in each one and the amount.

25 lb \$0.80 + \$1.20 = \$1.10

Again we must represent the *amount* in each mixture. The question asks, "How much of the \$1.20 coffee must be added?" So we let x equal the amount of the \$1.20 and $x + 25$ will be the amount in the total mixture.

Let x = pounds of \$1.20 coffee

25 lb \$0.80 + x lb \$1.20 = $(x + 25)$ lb \$1.10

2. The problem states that the *value* of the final mixture is equal to the *total* value of the other two. If we have the

number of pounds and the price per pound we can multiply to find the value of each mixture.

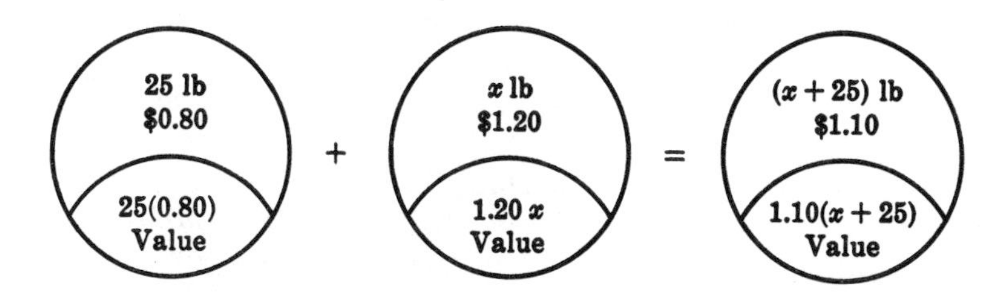

3. The product in the lower part of each circle represents the value of the mixture. The sum of the first two values equals the value of the total.

Equation

$$25(0.80) + 1.20\,x = 1.10(x + 25)$$

$$20 + 1.20\,x = 1.10\,x + 27.5$$

Multiply by 10 to clear the decimals. (You will still be correct if you multiply by 100.)

$$200 + 12x = 11x + 275$$

$$x = 75 \quad \textit{Answer}$$

Note that in this type of mixture problem the amounts add up to the total amount just as in the first type. And even though the problems are different, the basic procedure is the same. In one we have an amount and a percent. In the second, we have an amount and a price. And in both types, we find the products to set up the equation.

EXAMPLE 4. A girl scout troop has 20 pounds of candy worth 80 cents per pound. It wishes to mix it with candy worth 50 cents per pound so that the total mixture can be sold at 60 cents per pound without any gain or loss. How much of the 50-cent candy must be used?

Steps

1. Draw a diagram and put in prices and amounts. This problem is set up just like the percent problem except you have *cents* instead of *percent*.

   Let x = pounds of 50-cent candy

20 lb, 80¢ + x lb, 50¢ = $(x + 20)$ lb, 60¢

2. Multiply the cents per pound times the amounts. This will give you the *value* of each mixture. The *value of each original mixture added together should equal the value of the total mixture* if there is no gain or loss. (We assume no gain or loss unless the problem states otherwise.)

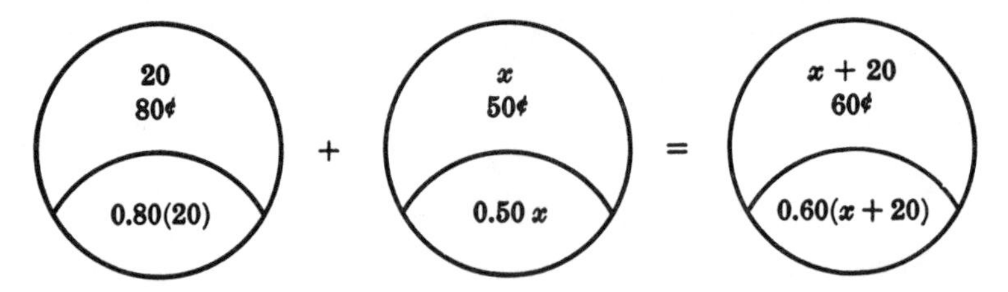

Remember, 80¢ should be written as a decimal when multiplying (0.80).

Equation

$$\begin{aligned} 0.80(20) + 0.50\,x &= 0.60(x + 20) \\ 16 + 0.50\,x &= 0.60\,x + 12 \\ 1600 + 50x &= 60x + 1200 \\ 50x - 60x &= 1200 - 1600 \\ -10x &= -400 \\ x &= 40 \quad \textit{Answer} \end{aligned}$$

Note: You could clear decimals in this equation by multiplying by 10 instead of 100. It would make all numbers in the equation whole numbers.

Unless otherwise stated, as we said, it is assumed that the value of a total mixture is equal to the sums of the values of the two original mixtures. But you *could* be given other relationships, as for instance, in the following problem.

EXAMPLE 5. A merchant purchased two lots of shoes. One lot he purchased for \$8 per pair and the second lot he purchased for \$10 per pair. There were 50 pairs in the first lot. How many pairs were in the second lot if he sold them all at $15 per pair and made a gain of $700 on the entire transaction?

Steps

1. Draw a diagram, fill in all information, and multiply price times quantity to get total value of each original lot and value of total lot.

Let x = pairs in second lot

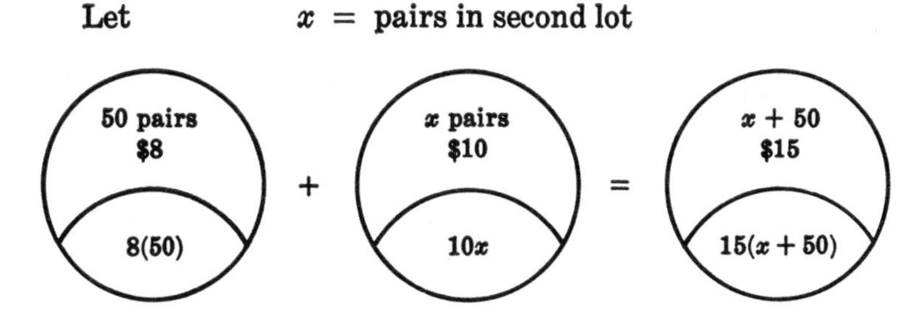

2. Set up the equation by adding $700 to the sum of the first and second lots. This addition of $700 equals the gain on the entire transaction.

Equation

$$8(50) + 10x + 700 = 15(x + 50)$$
$$400 + 10x + 700 = 15x + 750$$
$$-5x = -350$$
$$x = 70 \quad \textit{Answer}$$

Check

$$400 + 10(70) + 700 = 15(70) + 750$$
$$400 + 700 + 700 = 1050 + 750$$
$$1800 = 1800$$

There are some mixture problems in which water is added to dilute a mixture. Water has 0% acid, of course. You may also add pure ingredients to strengthen a mixture. For example, you might add pure sulphuric acid to a 30% solution to strengthen it. The pure ingredient is 100%. Remember, $0\%\, x = 0$, and $100\%\, x = x$. When a percent is changed to a decimal, the decimal is moved two places left.

Percent		Decimal
100%	=	1.00
80%	=	0.80
70%	=	0.70

EXAMPLE 6. A chemist needs a solution of tannic acid 70% pure. How much distilled water must he add to 5 gallons of acid which was 90% pure to obtain the 70% solution?

Solution

Let x = gallons of water

Diagram:

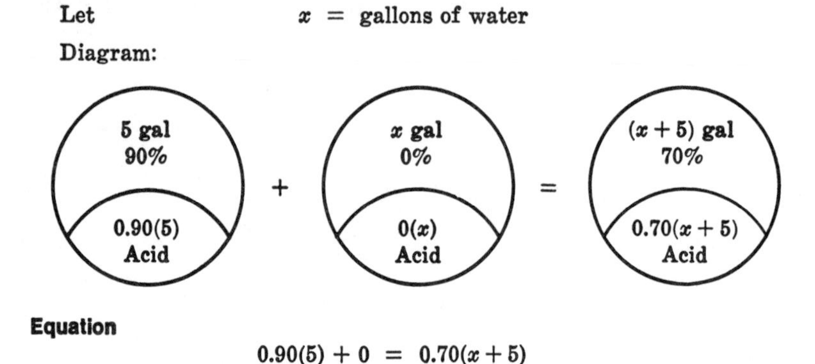

Equation

$$0.90(5) + 0 = 0.70(x + 5)$$
$$4.5 + 0 = 0.70\,x + 3.5$$

Multiply by 10.

$$45 = 7x + 35$$
$$-7x = 35 - 45$$
$$-7x = -10$$
$$x = 1\tfrac{3}{7} \quad \textit{Answer}$$

EXAMPLE 7. How much pure alcohol must a nurse add to 10 cc of a 60% alcohol solution to strengthen it to a 90% solution?

Solution

Let x = cc of pure alcohol

Diagram:

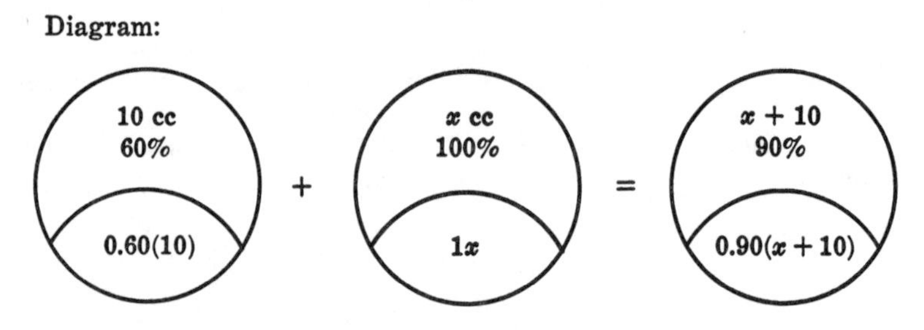

Pure alcohol is 100% alcohol. In decimal form 100% is 1.00 or 1.

Equation

$$
\begin{aligned}
0.60(10) + 1x &= 0.90(x + 10)\\
6 + x &= 0.90\,x + 9\\
60 + 10x &= 9x + 90\\
x &= 30 \quad \textit{Answer}
\end{aligned}
$$

Check

$$
\begin{aligned}
0.60(10) + 30 &= 0.90(30 + 10)\\
6 + 30 &= 27 + 9\\
36 &= 36
\end{aligned}
$$

Now that you understand a basic mixture problem, we can try a different kind. In this type of problem we take some from, and add some to, a given mixture. (You are more likely to find this type in second-year algebra.)

EXAMPLE 8. A 5-gallon radiator containing a mixture of water and antifreeze was supposed to contain a 50% antifreeze solution. When tested, it was found to have only 40% antifreeze. How much must be drained out and replaced with pure antifreeze so that the radiator will then contain the desired 50% antifreeze solution?

Solution

This time we need another circle for the diagram. The first circle in the diagram shows what the radiator has to start with. The second circle shows what has been drained off. It will always be the *same percent* as the solution from which it is taken. The third circle shows what is put back. It will be the *same amount* as that taken out. The fourth circle represents the final solution and will be the same amount you started with. You then multiply percent times amount to get amount of pure antifreeze in each solution and use the *multiples* to set up the equation, the same as before.

Let x = gallons of solution drained *and* replaced

Diagram:

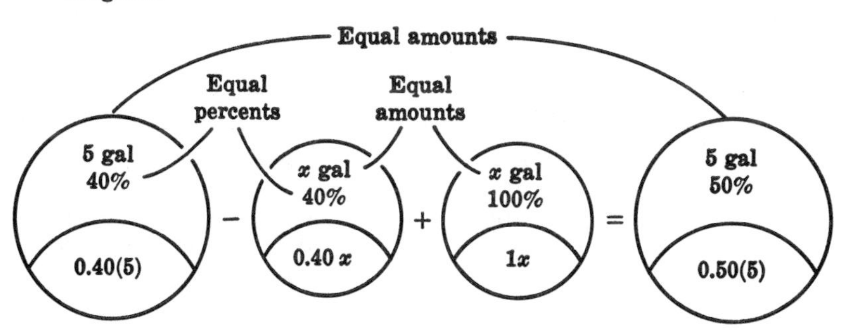

Note that you start with 5 gallons and finish with 5 gallons. The amount you drain off is 40% antifreeze, the same percent as in the original solution.

Equation

$$\begin{aligned} 0.40(5) - 0.40\,x + 1x &= 0.50(5) \\ 2.00 - 0.40\,x + 1x &= 2.50 \\ 200 - 40x + 100x &= 250 \\ 60x &= 50 \\ x &= \frac{5}{6} \quad \textit{Answer} \end{aligned}$$

EXAMPLE 9. A chemist has 300 grams of 20% hydrochloric acid solution. He wishes to drain some off and replace it with an 80% solution so as to obtain a 25% solution. How many grams must he drain and replace with the 80% solution?

Let x = grams of acid to drain and replace

Diagram:

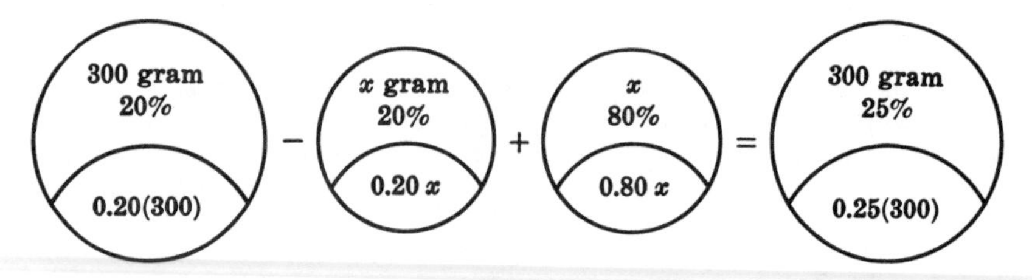

Equation

$$\begin{aligned} 0.20(300) - 0.20\,x + 0.80\,x &= 0.25(300) \\ 60 - 0.20\,x + 0.80\,x &= 75 \\ 6000 - 20x + 80x &= 7500 \\ 60x &= 1500 \\ x &= 25 \quad \textit{Answer} \end{aligned}$$

SUPPLEMENTARY MIXTURE PROBLEMS

1. A florist wishes to make bouquets of mixed spring flowers. Each bouquet is to be made up of tulips at $10 a bunch and daffodils at $7 a bunch. How many bunches of each should she use to make 15 bunches which she can sell for $8 a bunch?

2. A farmer has 100 gallons of 70% pure disinfectant. He wishes to mix it with disinfectant which is 90% pure in order to obtain 75% pure disinfectant. How much of the 90% pure must he use?

3. If an alloy containing 30% silver is mixed with a 55% silver alloy to get 800 pounds of 40% alloy, how much of each must be used?

4. Russ discovers at the end of the summer that his radiator antifreeze solution has dropped below the safe level. If the radiator contains 4 gallons of a 25% solution, how many gallons of pure antifreeze must he add to bring it up to a desired 50% solution?

5. A store manager wishes to reduce the price on his fresh ground coffee by mixing two grades. If he has 50 pounds of coffee which sells for $5 per pound, how much coffee worth $3 per pound must he mix with it so that he can sell the final mixture for $4.25 per pound?

6. A hospital needs to dilute a 50% boric acid solution to a 10% solution. If it needs 25 liters of the 10% solution, how much of the 50% solution and how much water should it use?

7. Annabell has 25 ounces of a 20% boric acid solution which she wishes to dilute to a 10% solution. How much water does she have to add in order to obtain the 10% solution?

8. Forty liters of a 60% disinfectant solution are to be mixed with a 10% solution to dilute it to a 20% solution. How much of the 10% solution must be used?

9. A doctor orders 20 grams of a 52% solution of a certain medicine. The pharmacist has only bottles of 40% and bottles of 70% solution. How much of each must he use to obtain the 20 grams of the 52% solution?

10. Forty liters of a 60% salt solution are reduced to a 45% solution. How much must be drained off and replaced with distilled water so that the resulting solution will contain only 45% salt?

11. A pharmacist has to fill a prescription calling for 5 ounces of a 12% argyrol solution. She has 5 ounces of a 15% solution and 5 ounces of a 5% solution. If she starts with the 5 ounces of 15%, how much must she draw off and replace with 5% in order to fill the prescription?

SOLUTIONS TO SUPPLEMENTARY MIXTURE PROBLEMS

1. Let x = number of \$10 bunches

$15 - x$ = number of \$7 bunches

Diagram:

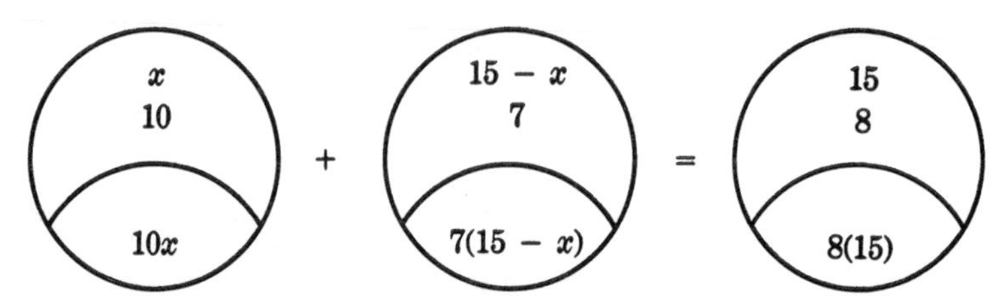

The value of the bunches at \$10 each is $10x$. There are 15 bunches in total so $15 - x$ is the number of bunches at \$7 each. The value of the bunches in the second mixture is $7(15 - x)$. Value of the first mixture plus the value of the second mixture equals the value of the total mixture.

Equation

$$10x + 7(15 - x) = 8(15)$$

$$10x + 105 - 7x = 120$$

$$3x = 15$$

$x = 5$ bunches at \$10 each
$15 - x = 10$ bunches of \$7 each
Answers

Check

$$10(5) + 7(15 - 5) = 8(15)$$

$$50 + 105 - 35 = 120$$

$$120 = 120$$

2. Let x = gallons of 90% disinfectant

Diagram:

70% of 100 gallons equals pure disinfectant in first mixture. 90% of x gallons equals pure disinfectant in second mixture. 75% of $(x+100)$ gallons equals pure disinfectant in total mixture.

Equation

$$0.70(100) + 0.90\,x = 0.75(x+100)$$
$$70 + 0.90\,x = 0.75\,x + 75$$

Multiply by 100.

$$7000 + 90x = 75x + 7500$$
$$15x = 500$$
$$x = 33\tfrac{1}{3} \quad \textit{Answer}$$

Check

$$0.70(100) + 0.90(33\tfrac{1}{3}) = 0.75(33\tfrac{1}{3} + 100)$$
$$70 + 30 = 25 + 75$$
$$100 = 100$$

3. Let

$$x = \text{lbs of 30\% alloy}$$
$$800 - x = \text{lbs of 55\% alloy}$$

Diagram:

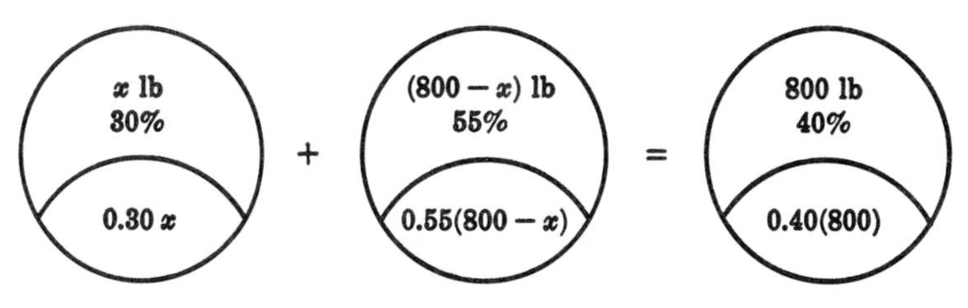

Equation

$$0.30\,x + 0.55(800 - x) = 0.40(800)$$

Clear parentheses.

$$0.30\,x + 440 - 0.55\,x = 320$$

Multiply by 100.

$$30x + 44{,}000 - 55x = 32{,}000$$
$$-25x = -12{,}000$$
$$\left.\begin{aligned} x &= 480 \\ 800 - x &= 320 \end{aligned}\right\} \quad \textit{Answers}$$

Check

$$0.30(480) + 0.55(800 - 480) = 0.40(800)$$
$$144 + 440 - 264 = 320$$
$$584 - 264 = 320$$
$$320 = 320$$

4. Let x = gallons of pure antifreeze

Diagram:

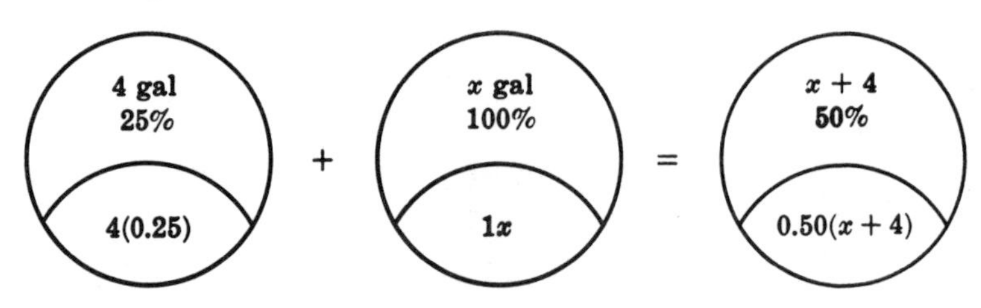

Equation

$$0.25(4) + 1x = 0.50(x + 4)$$
$$1.00 + x = 0.50\,x + 2.00$$

Multiply by 100.

$$100 + 100x = 50x + 200$$
$$x = 2 \quad \textit{Answer}$$

Check

$$0.25(4) + 2 = 0.50(2 + 4)$$
$$1 + 2 = 1 + 2$$

5. Let x = pounds of \$3 coffee

Diagram:

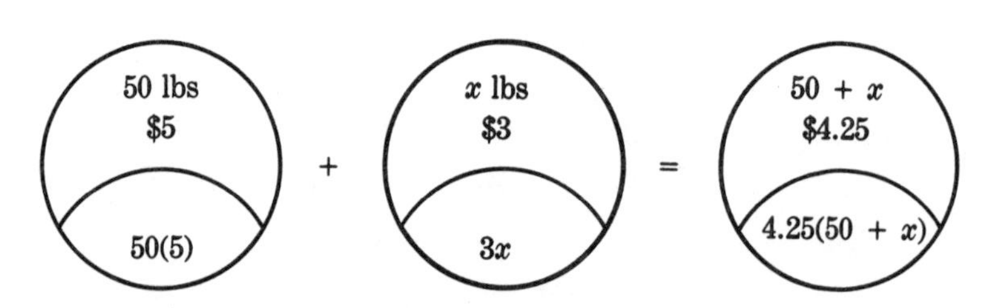

Equation

$$50(5) + 3x = 4.25(50 + x)$$
$$250 + 3x = 212.50 + 4.25x$$
$$25{,}000 + 300x = 21{,}250 + 425x$$
$$-125x = -3{,}750$$
$$x = 30 \text{ pounds of \$3 coffee} \qquad \textit{Answer}$$

6. Let

$$x = \text{liters of 50\% solution}$$
$$25 - x = \text{liters of water}$$

Diagram:

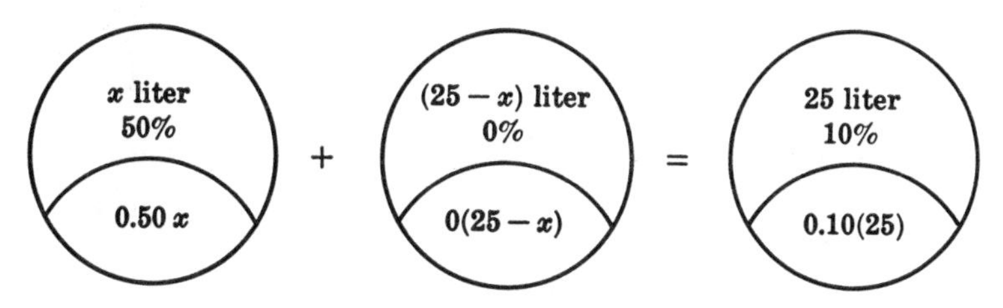

Equation

$$0.50x + 0 = 0.10(25)$$
$$0.50x = 2.50$$
$$50x = 250$$
$$\left.\begin{aligned} x &= 5 \\ 25 - x &= 20 \end{aligned}\right\} \quad \textit{Answers}$$

7. Let

$$x = \text{ounces of water needed}$$

Diagram:

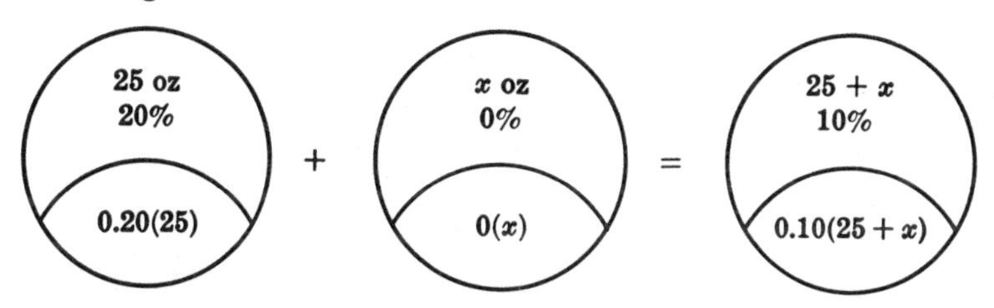

Water has 0% boric acid in it. The percent times the amount in each mixture gives the number of ounces of pure boric acid in each. This amount of pure acid is represented at the bottom of each circle.

Equation

$$0.20(25) + 0(x) = 0.10(25 + x)$$
$$5 + 0 = 2.5 + 0.1\,x$$

Multiply by 10 to clear fractions.

$$50 = 25 + x$$
$$x = 25 \quad \textit{Answer}$$

8. Let x = liters of 10% solution

Diagram:

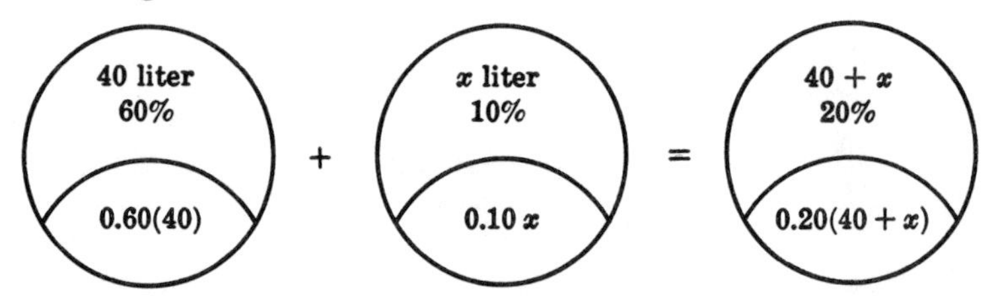

Equation

$$0.60(40) + 0.10\,x = 0.20(40 + x)$$
$$24 + 0.10\,x = 8 + 0.20\,x$$

Multiply by 10 to clear decimals.

$$240 + 1x = 80 + 2x$$
$$-x = -160$$
$$x = 160 \quad \textit{Answer}$$

Check

$$0.60(40) + 0.10(160) = 0.20(40 + 160)$$
$$24 + 16 = 0.20(200)$$
$$40 = 40$$

9. Let x = grams of 40% solution

$20 - x$ = grams of 70% solution

Diagram:

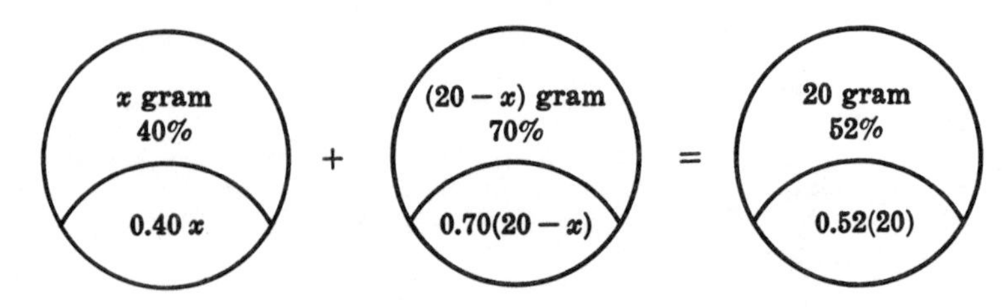

Equation

$$0.40x + 0.70(20 - x) = 0.52(20)$$

Clear parentheses.

$$0.40x + 14 - 0.70x = 10.4$$

Clear decimals.

$$4x + 140 - 7x = 104$$

$$-3x = -36$$

$$\left.\begin{aligned} x &= 12 \\ 20 - x &= 8 \end{aligned}\right\} \quad \textit{Answers}$$

Check

$$0.40(12) + 0.70(20 - 12) = 0.52(20)$$

$$4.8 + 14 - 8.4 = 10.4$$

$$18.8 - 8.4 = 10.4$$

$$10.4 = 10.4$$

10. Let x = liters of distilled water

Diagram:

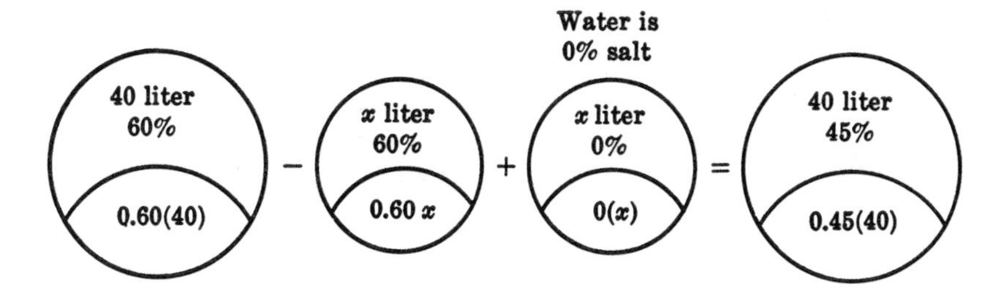

Equation

$$0.60(40) - 0.60x + 0(x) = 0.45(40)$$

$$24 - 0.60x = 18$$

$$2400 - 60x = 1800$$

$$-60x = -600$$

$$x = 10 \quad \textit{Answer}$$

11. Let x = ounces of 5% solution

Diagram:

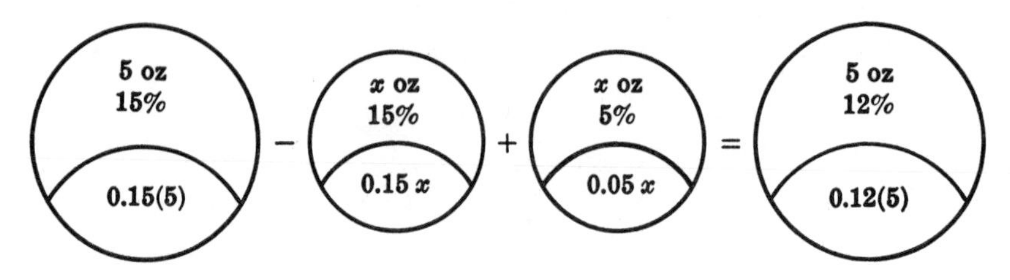

Equation

$$
\begin{aligned}
0.15(5) - 0.15x + 0.05x &= 0.12(5) \\
0.75 - 0.15x + 0.05x &= 0.60 \\
75 - 15x + 5x &= 60 \\
-10x &= -15 \\
x &= 1\tfrac{1}{2} \quad \textit{Answer}
\end{aligned}
$$

Chapter 4

Coins

Problems about money, either metal or paper, stamps, or any objects of value, are, for want of a better classification, grouped under the heading "Coin Problems." It is very important to identify and label whether it is *how many things* or *how much money* you are representing.

EXAMPLE 1. Michael has some coins in his pocket consisting of dimes, nickels, and pennies. He has two more nickels than dimes, and three times as many pennies as nickels. How many of each kind of coin does he have if the total value is 52¢?

Steps

1. Try to determine which coin he has *fewest of*. This is often a good way to find what to let x stand for. Here he has fewer dimes than nickels or pennies.

2. The question asks *how many* of each kind of coin (not how much they are *worth*!). That is, what *number* of each kind of coin does he have? So, let

$$x = \text{number of dimes}$$

3. Go back to one fact at a time. He has two more nickels than dimes.

$$x + 2 = \text{number of nickels}$$

4. Next fact, "three times as many pennies as nickels."

$$3(x+2) = \text{number of pennies}$$

5. The information left for the equation is that the total *value* is 52¢. You cannot say that the total *number of coins* equals 52¢. *Number of* must be changed to *value.* If you had two quarters you would have 50¢. You multiplied how many coins by how much each is worth. So let's change

our *how many* to *how much.* If you have the number of dimes, multiply by 10 to change to cents. Multiply the number of nickels by 5 to change to cents. The number of pennies is the same as the number of cents.

Number of Coins	Value in Cents
x = number of dimes	$10x$ = number of cents in dimes
$x + 2$ = number of nickels	$5(x + 2)$ = number of cents in nickels
$3(x + 2)$ = number of pennies	$3(x + 2)$ = number of cents in pennies

Now we can add the *amounts of money.* If you make it all pennies, there are no decimals.

$$10x + 5(x + 2) + 3(x + 2) = 52$$
$$10x + 5x + 10 + 3x + 6 = 52$$
$$18x = 36$$
$$\left.\begin{aligned} x &= 2 \\ x + 2 &= 4 \\ 3(x + 2) &= 12 \end{aligned}\right\} \quad \textit{Answers}$$

Check

2(10¢)	=	20 cents in dimes
4(5¢)	=	20 cents in nickels
12(1¢)	=	12 cents in pennies
Total		52 cents

EXAMPLE 2. A coin collector had a collection of silver coins worth $205. There were five times as many quarters as half dollars (50¢) and 200 fewer dimes than quarters. How many of each kind of coin did the collector have?

Solution

Number of Coins	Value in Cents
x = number of half dollars	$50x$ = amount in cents
$5x$ = number of quarters	$25(5x)$ = amount in cents
$5x - 200$ = number of dimes	$10(5x - 200)$ = amount in cents

Equation

Remember, everything has been changed to *cents.* So \$205 has to be changed to cents by multiplying by 100. \$205 = 20,500 cents.

$$
\begin{aligned}
50x + 25(5x) + 10(5x - 200) &= 20{,}500 \\
50x + 125x + 50x - 2000 &= 20{,}500 \\
225x &= 22{,}500 \\
x &= 100 \\
5x &= 500 \\
5x - 200 &= 300
\end{aligned}
\quad \text{Answers}
$$

Check

$$
\begin{aligned}
100(50¢) &= \$\ 50 \\
500(25¢) &= \ \ 125 \\
300(10¢) &= \ \ \ \underline{30} \\
\text{Total} &\ \ \$205
\end{aligned}
$$

EXAMPLE 3. Mr. Dinkelspiel bought \$30.64 worth of stamps. He bought 20 more 19 cent stamps than 50 cent stamps. He bought twice as many 29 cent stamps as 19 cent stamps. How many of each kind of stamp did he buy?

Solution

Number of Stamps	Value in Cents
x = number of 50 cent stamps	$50x$
$x + 20$ = number of 19 cent stamps	$19(x + 20)$
$2(x + 20)$ = number of 29 cent stamps	$29[2(x + 20)]$

Equation

Remember, \$30.64 equals 3064 cents.

$$
\begin{aligned}
50x + 19(x + 20) + 29[2(x + 20)] &= 3064 \\
50x + 19x + 380 + 58x + 1160 &= 3064 \\
127x &= 1524 \\
x &= 12 \\
x + 20 &= 32 \\
2(x + 20) &= 64
\end{aligned}
\quad \text{Answers}
$$

Check

$$12(50) + 19(32) + 64(29) = 3064$$
$$600 + 608 + 1856 = 3064$$
$$3064 = 3064$$

The principle of converting "how many things" into "how much money" can be applied to many types of problems. Let's try some.

SUPPLEMENTARY COIN PROBLEMS

1. A collection of coins has a value of 64¢. There are two more nickels than dimes and three times as many pennies as dimes. How many of each kind of coin are there?

2. Tanya has ten bills in her wallet. She has a total of $40. If she has one more $5 bill than $10 bills, and two more $1 bills than $5 bills, how many of each does she have? (There are two ways of working this problem. See if you can do it both ways.)

3. Mario bought $20 worth of stamps at the post office. He bought ten more 4 cent stamps than 19 cent stamps. The number of 29 cent stamps was three times the number of 19 cent stamps. He also bought two $1 stamps. How many of each kind of stamp did he purchase?

4. Mr. Abernathy purchases a selection of wrenches for his shop. His bill is $78. He buys the same number of $1.50 and $2.50 wrenches, and half that many $4 wrenches. The number of $3 wrenches is one more than the number of $4 wrenches. How many of each did he purchase? (Hint: If you have not worked with fractions, use decimals for all fractional parts.)

5. A clerk at the Dior Department Store receives $15 in change for her cash drawer at the start of each day. She receives twice as many dimes as fifty-cent pieces, and the same number of quarters as dimes. She has twice as many nickels as dimes and a dollar's worth of pennies. How many of each kind of coin does she receive?

6. A collection of 36 coins consists of nickels, dimes, and quarters. There are three fewer quarters than nickels and six more dimes than quarters. How many of each kind of coin is there?

7. The cash drawer of the market contains $227 in bills. There are six more $5 bills than $10 bills. The number of $1 bills is two more than 24 times the number of $10 bills. How many bills of each kind are there?

8. Jennifer went to the drugstore for cosmetics and sundries. She bought a bottle of aspirin and a bottle of Tylenol. The aspirin cost $1.25 more than the Tylenol. She also bought cologne which cost twice as much as the total of the other two combined. How much did each cost if her total (without tax) was $24.75?

9. Terry bought some gum and some candy. The number of packages of gum was one more than the number of mints. The number of mints was three times the number of candy bars. If gum was 24 cents a package, mints were 10 cents each, and candy bars were 35 cents each, how many of each did he get for $5.72?

10. Jose Ramirez had $35 to buy his groceries. He needed milk at $1.30 a carton, bread at $1.69 a loaf, breakfast cereal at $2 a box, and meat at $3.59 a pound. He bought twice as many cartons of milk as loaves of bread. The number of packages of cereal was one more than the number of loaves of bread. The number of pounds of meat was the same as the number of packages of cereal. How many of each did he purchase if he received $9.65 in change?

SOLUTIONS TO SUPPLEMENTARY COIN PROBLEMS

1.

Number of Coins	Value in Cents
x = number of dimes	$10x$ = value of dimes
$x + 2$ = number of nickels	$5(x + 2)$ = value of nickels
$3x$ = number of pennies	$3x$ = value of pennies

The total value of the dimes, nickels, and quarters equals 64 cents.

Equation

$$10x + 5(x+2) + 3x = 64$$
$$10x + 5x + 10 + 3x = 64$$
$$18x = 54$$
$$\left.\begin{aligned} x &= 3 \\ x+2 &= 5 \\ 3x &= 9 \end{aligned}\right\} \textit{Answers}$$

Check

$$\begin{aligned} 3(10¢) &= 30¢ \\ 5(5¢) &= 25 \\ 9(1¢) &= 9 \\ \text{Total} &\ \ 64¢ \end{aligned}$$

2.

Number of Bills		Value in Dollars	
x	= number of \$10 bills	$10x$	= value of \$10 bills
$x+1$	= number of \$5 bills	$5(x+1)$	= value of \$5 bills
$(x+1)+2$	= number of \$1 bills	$1(x+3)$	= value of \$1 bills

Equation

$$10x + 5(x+1) + (x+3) = 40$$
$$10x + 5x + 5 + x + 3 = 40$$
$$16x = 32$$
$$\left.\begin{aligned} x &= 2 \\ x+1 &= 3 \\ x+3 &= 5 \end{aligned}\right\} \textit{Answers}$$

Check

$$\begin{aligned} 2(\$10) &= \$20 \\ 3(\$5) &= 15 \\ 5(\$1) &= 5 \\ \text{Total} &\ \ \$40 \end{aligned}$$

Alternate Solution

The information that there are 10 bills was not used in the first solution. *If* you know *how many* coins or bills or stamps you have, the problem is simplified because you do not have to convert to money.

Let

$$x = \text{number of \$10 bills}$$
$$x + 1 = \text{number of \$5 bills}$$
$$x + 3 = \text{number of \$1 bills}$$

Equation

$$x + (x + 1) + (x + 3) = 10$$
$$3x + 4 = 10$$
$$3x = 6$$
$$\left.\begin{aligned} x &= 2 \\ x + 1 &= 3 \\ x + 3 &= 5 \end{aligned}\right\} \quad \textit{Answers}$$

Remember, however, that you are seldom given the *number* of coins.

3.

Number of Stamps	Value in Cents
Let x = no. of 19 cent stamps	$19x$
$x + 10$ = no. of 4 cent stamps	$4(x + 10)$
$3x$ = no. of 29 cent stamps	$29(3x)$
2 = no. of \$1 stamps	$2(100)$

Equation

$$19x + 4(x + 10) + 29(3x) + 2(100) = 2000$$
$$19x + 4x + 40 + 87x + 200 = 2000$$
$$110x = 1760$$
$$\left.\begin{aligned} x &= 16 \\ x + 10 &= 26 \\ 3x &= 48 \end{aligned}\right\} \quad \textit{Answers}$$

Check

$$19(16) + 4(26) + 29(48) + 200 = 2000$$
$$304 + 104 + 1392 + 200 = 2000$$
$$2000 = 2000$$

4. Number of Wrenches

$$x = \text{number of \$4.00 wrenches}$$
$$2x = \text{number of \$1.50 wrenches}$$
$$2x = \text{number of \$2.50 wrenches}$$
$$x + 1 = \text{number of \$3.00 wrenches}$$

Value in Dollars

$$4x = \text{value of \$4.00 wrenches}$$
$$2x(1.50) = \text{value of \$1.50 wrenches}$$
$$2x(2.50) = \text{value of \$2.50 wrenches}$$
$$3(x + 1) = \text{value of \$3.00 wrenches}$$

Equation

$$4x + 2x(1.50) + 2x(2.50) + 3(x + 1) = 78$$

Note in this problem that the decimal is removed when you clear parentheses.

$$4x + 3x + 5x + 3x + 3 = 78$$
$$15x = 75$$
$$\left.\begin{aligned} x &= 5 \\ 2x &= 10 \\ 2x &= 10 \\ x + 1 &= 6 \end{aligned}\right\} \textit{Answers}$$

Check

$$5(\$4.00) = \$20$$
$$10(\$1.50) = 15$$
$$10(\$2.50) = 25$$
$$6(\$3.00) = \underline{18}$$
$$\text{Total } \$78$$

5. Let

Number of Coins

$$x = \text{number of 50¢ pieces}$$
$$2x = \text{number of 10¢ pieces}$$
$$2x = \text{number of 25¢ pieces}$$
$$4x = \text{number of 5¢ pieces}$$
$$100 = \text{number of 1¢ pieces}$$

Value in Cents

$$50x = \text{value of 50¢ pieces}$$
$$10(2x) = \text{value of 10¢ pieces}$$
$$25(2x) = \text{value of 25¢ pieces}$$
$$5(4x) = \text{value of 5¢ pieces}$$
$$1(100) = \text{value of 1¢ pieces}$$

Equation

Fifteen dollars equals 1500 cents.

$$50x + 2x(10) + 2x(25) + 4x(5) + 100 = 1500$$
$$50x + 20x + 50x + 20x + 100 = 1500$$
$$140x = 1400$$
$$\left.\begin{aligned} x &= 10 \\ 2x &= 20 \\ 2x &= 20 \\ 4x &= 40 \end{aligned}\right\} \textit{Answers}$$

Check

$$\begin{aligned} 10(50¢) &= \$\ 5 \\ 20(10¢) &= \ \ 2 \\ 20(25¢) &= \ \ 5 \\ 40(5¢) &= \ \ 2 \\ 100(1¢) &= \ \ \underline{1} \\ \text{Total} & \ \ \$15 \end{aligned}$$

6. Let

x = number of quarters (there are *fewer* quarters)

$x + 3$ = number of nickels

$x + 6$ = number of dimes

Here we do *not* change to cents because the *number* of coins is given.

Equation

$$x + (x + 3) + (x + 6) = 36$$
$$3x + 9 = 36$$
$$3x = 27$$

Answers

$$x = 9 \text{ quarters}$$
$$x + 3 = 12 \text{ nickels}$$
$$x + 6 = 15 \text{ dimes}$$

Check

$$9 + 12 + 15 = 64$$
$$64 = 64$$

7. Let

$$x = \text{number of \$10 bills}$$
$$x + 6 = \text{number of \$5 bills}$$
$$24x + 2 = \text{number of \$1 bills}$$

Change all to amount of money in dollars.

$$10x = \text{number of dollars in \$10 bills}$$
$$5(x+6) = \text{number of dollars in \$5 bills}$$
$$24x + 2 = \text{number of dollars in \$1 bills}$$

Equation

$$10x + 5(x+6) + 24x + 2 = 227$$
$$10x + 5x + 30 + 24x + 2 = 227$$
$$39x + 32 = 227$$
$$39x = 195$$
$$\left.\begin{aligned} x &= 5 \\ x + 6 &= 11 \\ 24x + 2 &= 122 \end{aligned}\right\} \textit{Answers}$$

Check

$$10(5) + 5(11) + 24(5) + 2 = 227$$
$$50 + 55 + 120 + 2 = 227$$
$$227 = 227$$

8. Let

$$\text{Let } x = \text{cost in cents of Tylenol}$$
$$x + 125 = \text{cost in cents of aspirin}$$
$$2(2x + 125) = \text{cost in cents of cologne}$$

Equation

$$x + x + 125 + 2(2x + 125) = 2475$$
$$2x + 125 + 4x + 250 = 2475$$
$$6x + 375 = 2475$$
$$6x = 2100$$
$$\left.\begin{aligned} x &= 350 \\ x + 125 &= 475 \\ 2(2x + 125) &= 1650 \end{aligned}\right\} \textit{Answers}$$

Check

$$3.50 + 4.75 + 16.50 = 24.75$$
$$24.75 = 24.75$$

9.

Let x = no. of candy bars @ 35 cents each

$3x$ = no. of mints @ 10 cents each

$3x + 1$ = no. of packages of gum @ 24 cents each

Value in Cents

$35x$ = cents for candy bar

$10(3x)$ = cents for mints

$24(3x + 1)$ = cents for gum

Equation

$$35x + 10(3x) + 24(3x + 1) = 572$$
$$35x + 30x + 72x + 24 = 572$$
$$137x + 24 = 572$$
$$137x = 548$$
$$\left.\begin{aligned} x &= 4 \\ 3x &= 12 \\ 3x + 1 &= 13 \end{aligned}\right\} \textit{Answers}$$

Check

$$4(35) + 12(10) + 13(24) = 572$$
$$140 + 120 + 312 = 572$$
$$572 = 572$$

10.

Let x = no. of loaves of bread @ \$1.69 each

$2x$ = no. of cartons of milk @ \$1.30

$x + 1$ = no. of packages of cereal @ \$2.00

$x + 1$ = no. of pounds of meat @ \$3.59

Cost in Cents

$$169x = \text{cost of bread}$$

$$130(2x) = \text{cost of milk}$$

$$200(x + 1) = \text{cost of cereal}$$

$$359(x + 1) = \text{cost of meat}$$

Equation

$$169x + 130(2x) + 200(x + 1) + 359(x + 1) = 3500 - 965$$

$$169x + 260x + 200x + 200 + 359x + 359 = 2535$$

$$988x = 1976$$

$$\left.\begin{aligned} x &= 2 \\ 2x &= 4 \\ x + 1 &= 3 \\ x + 1 &= 3 \end{aligned}\right\} \textit{Answers}$$

Check

$$169(2) + 130(4) + 200(3) + 359(3) = 3500 - 965$$

$$338 + 520 + 600 + 1077 = 2535$$

$$2535 = 2535$$

Chapter 5

Age

Age problems usually follow a certain pattern. That is, they usually have certain basic facts. Frequently, they refer to ages at different points in time.

For example, Ted's father is twice as old as Ted. This means that their ages *now* could be represented

$$x = \text{Ted's age now}$$

$$2x = \text{father's age now}$$

If the problem then states that 10 years ago the father was three times as old, *you have to represent their ages 10 years ago.* To show their ages 10 years ago, you don't need any other information. All you do is subtract 10 from their ages now. (If you know your age *now,* you subtract 10 to find your age 10 years ago.)

$$x - 10 = \text{Ted's age 10 years ago}$$

$$2x - 10 = \text{father's age 10 years ago}$$

Now you can use the fact that the father was *then* three times as old as Ted to set up the equation.

Father's age 10 years ago was three times Ted's 10 years ago.

$$2x - 10 = 3(x - 10)$$

This procedure of representing ages now, moving them forward or backward, and then setting up the equation, is often used in age problems.

EXAMPLE 1. Mary's father is four times as old as Mary. Five years ago he was seven times as old. How old is each now?

Solution

First fact: Mary's father is four times as old as Mary.

Let x = Mary's age *now* (smaller number)

$4x$ = father's age *now*

Second fact: Five years ago...(stop!) The problem tells you about their ages at *another time*. Five years ago your age would be 5 less than your age now. So

$x - 5$ = Mary's age 5 years ago

$4x - 5$ = father's age 5 years ago

Now read on. Five years ago he was *seven times as old* (as she was).

$$4x - 5 = 7(x - 5)$$

$$4x - 5 = 7x - 35$$

$$-3x = -30$$

$$\left.\begin{aligned} x &= 10 \\ 4x &= 40 \end{aligned}\right\} \quad \textit{Answers}$$

Check

$$\left.\begin{aligned} x - 5 &= 5 \\ 4x - 5 &= 35 \end{aligned}\right\} \quad \text{and} \quad 7 \times 5 = 35$$

EXAMPLE 2. Abigail is 6 years older than Jonathan. Six years ago she was twice as old as him. How old is each now?

Solution

Let x = Jonathan's age now (smaller number)

$x + 6$ = Abigail's age now

$x - 6$ = Jonathan's age 6 years ago

$(x + 6) - 6$ = Abigail's age 6 years ago

Six years ago she was twice as old as him.

Equation

$$(x + 6) - 6 = 2(x - 6)$$

$$x = 2x - 12$$

$$-x = -12$$

$$\left.\begin{aligned} x &= 12 \\ x + 6 &= 18 \end{aligned}\right\} \quad \textit{Answers}$$

Note that in this type of problem, as soon as you come to the statements about a time other than now, stop and move the ages *now* forward or backward. "Three years ago" you subract 3 from ages *now*. "Five years from now" you add 5 to ages *now*. This is like saying that in 5 years you will be 5 years older. Be careful to use *only* the number of years you add or subtract and don't use any other information until the equation. The equation is usually based on the relationship in the future or past *after* you move the ages ahead or back.

EXAMPLE 3. Wolfgang's father is 26 years older than Wolfgang. In 10 years, the sum of their ages will be 80. What are their present ages?

Solution

Let

$$x = \text{Wolfgang's age now}$$
$$x + 26 = \text{father's age now}$$

$$x + 10 = \text{Wolfgang's age in 10 years}$$
$$(x + 26) + 10 = \text{father's age in 10 years}$$

Equation

$$x + 10 + (x + 26) + 10 = 80$$
$$2x + 46 = 80$$
$$2x = 34$$
$$\left.\begin{aligned} x &= 17 \\ x + 26 &= 43 \end{aligned}\right\} \quad \textit{Answers}$$

Check

In 10 years Wolfgang will be 27 and his father will be 53.

$$53 + 27 = 80$$

EXAMPLE 4. Mary is twice as old as Helen. If 8 is subtracted from Helen's age and 4 is added to Mary's age, Mary will then be four times as old as Helen. Find their ages.

Note: This is a simple number relationship problem.

Solution

Let

$$x = \text{Helen's age now}$$
$$2x = \text{Mary's age now}$$

$$x - 8 = 8 \text{ subtracted from Helen's age}$$
$$2x + 4 = 4 \text{ added to Mary's age}$$

Equation

$$2x + 4 = 4(x - 8)$$
$$-2x = -36$$
$$\left.\begin{aligned} x &= 18 \\ 2x &= 36 \end{aligned}\right\} \textit{Answers}$$

Check

$$2(18) + 4 = 4(18 - 8)$$
$$36 + 4 = 4(10)$$
$$40 = 40$$

SUPPLEMENTARY AGE PROBLEMS

1. A man is four times as old as his son. In 3 years, the father will be three times as old as the son. How old is each now?

2. Abigail is 8 years older than Cynthia. Twenty years ago Abigail was three times as old as Cynthia. How old is each now?

3. Seymour is twice as old as Cassandra. If 16 is added to Cassandra's age and 16 is subtracted from Seymour's age, their ages will then be equal. What are their present ages?

4. In 4 years Cranston's age will be the same as Terrill's age now. In 2 years, Terrill will be twice as old as Cranston. Find their ages now.

5. Mrs. Smythe is twice as old as her daughter Samantha. Ten years ago the sum of their ages was 46 years. How old is Mrs. Smythe?

6. A Roman statue is three times as old as a Florentine statue. One hundred years from now the Roman statue will be twice as old. How old is the Roman statue?

7. Sheri's age in 20 years will be the same as Terry's age is now. Ten years from now, Terry's age will be twice Sheri's. What are their present ages?

8. Bettina's age is three times Melvina's. If 20 is added to Melvina's age and 20 is subtracted from Bettina's, their ages will be equal. How old is each now?

SOLUTIONS TO SUPPLEMENTARY AGE PROBLEMS

1. Let

$$x = \text{son's age now}$$
$$4x = \text{father's age now}$$

$$x + 3 = \text{son's age in 3 years}$$
$$4x + 3 = \text{father's age in 3 years}$$

In 3 years, the father will be three times as old as the son.

Equation

$$4x + 3 = 3(x + 3)$$
$$4x + 3 = 3x + 9$$
$$\left.\begin{aligned} x &= 6 \\ 4x &= 24 \end{aligned}\right\} \textit{Answers}$$

2. Let

$$x = \text{Cynthia's age now}$$
$$x + 8 = \text{Abigail's age now}$$

$$x - 20 = \text{Cynthia's age 20 years ago}$$
$$(x + 8) - 20 = \text{Abigail's age 20 years ago}$$

Equation

$$(x + 8) - 20 = 3(x - 20)$$
$$x - 12 = 3x - 60$$
$$-2x = -48$$
$$\left.\begin{aligned} x &= 24 \\ x + 8 &= 32 \end{aligned}\right\} \textit{Answers}$$

3. Let

$$x = \text{Cassandra's age now}$$
$$2x = \text{Seymour's age now}$$

$$x + 16 = \text{16 added to Cassandra's age}$$
$$2x - 16 = \text{16 subtracted from Seymour's age}$$

Equation

$$2x - 16 = x + 16$$
$$\left.\begin{aligned} x &= 32 \\ 2x &= 64 \end{aligned}\right\} \textit{Answers}$$

4. Let

$$x = \text{Terrill's age now (and Cranston's age in 4 years)}$$
$$x - 4 = \text{Cranston's age now}$$

$$x + 2 = \text{Terrill's age in 2 years}$$
$$(x - 4) + 2 = \text{Cranston's age in 2 years}$$

Terrill will be twice as old as Cranston in 2 years.

Equation

$$x + 2 = 2[(x - 4) + 2]$$
$$x + 2 = 2x - 8 + 4$$
$$x + 2 = 2x - 4$$
$$\left.\begin{aligned} x &= 6 \\ x - 4 &= 2 \end{aligned}\right\} \textit{Answers}$$

5. Let

$$x = \text{Samantha's age now}$$
$$2x = \text{Mrs. Smythe's age now}$$

$$x - 10 = \text{Samantha's age 10 years ago}$$
$$2x - 10 = \text{Mrs. Smythe's age 10 years ago}$$

Equation

$$x - 10 + 2x - 10 = 46$$
$$3x - 20 = 46$$
$$3x = 66$$
$$\left.\begin{aligned} x &= 22 \\ 2x &= 44 \end{aligned}\right\} \textit{Answers}$$

6. Let

$$x = \text{age of Florentine statue now}$$
$$3x = \text{age of Roman statue now}$$

$$x + 100 = \text{age of Florentine statue in 100 years}$$
$$3x + 100 = \text{age of Roman statue in 100 years}$$

Equation

$$3x + 100 = 2(x + 100)$$
$$3x + 100 = 2x + 200$$
$$\left.\begin{aligned} x &= 100 \\ 3x &= 300 \end{aligned}\right\} \textit{Answers}$$

7. Let

$$
\begin{aligned}
x &= \text{Terry's age now} \\
x - 20 &= \text{Sheri's age now} \\
x + 10 &= \text{Terry's age in 10 years} \\
(x - 20) + 10 &= \text{Sheri's age in 10 years}
\end{aligned}
$$

Equation

$$
\begin{aligned}
x + 10 &= 2[(x - 20) + 10] \\
x + 10 &= 2(x - 10) \\
x + 10 &= 2x - 20 \\
-x &= -30 \\
x &= 30 \\
x + 20 &= 10
\end{aligned}
\quad \Bigg\} \; \textit{Answers}
$$

8. Let

$$
\begin{aligned}
x &= \text{Melvina's age now} \\
3x &= \text{Bettina's age now} \\
x + 20 &= \text{20 added to Melvina's age} \\
3x - 20 &= \text{20 subtracted from Bettina's age}
\end{aligned}
$$

Equation

$$
\begin{aligned}
x + 20 &= 3x - 20 \\
-2x &= -40 \\
x &= 20 \\
3x &= 60
\end{aligned}
\quad \Bigg\} \; \textit{Answers}
$$

Chapter 6

Levers

Most children are familiar with teeter-totters or see-saws. They know that if fat Sam sits on one end and skinny Jane sits on the other, Jane goes up in the air and Sam sinks to the ground. So Sam slides toward the center (called balance point or *fulcrum*) until they are in balance.

This example illustrates the fact that in order to balance, unequal weights (or forces) must be different distances from the point of balance. The teeter-totter, board, or rod, is called the *lever*. The principle involved is that the weight times its distance from the fulcrum on one side equals the weight times *its* distance from the fulcrum on the other side. If you let w_1 and d_1 represent one weight and distance, and w_2 and d_2 represent the others, you might diagram this relationship like this:

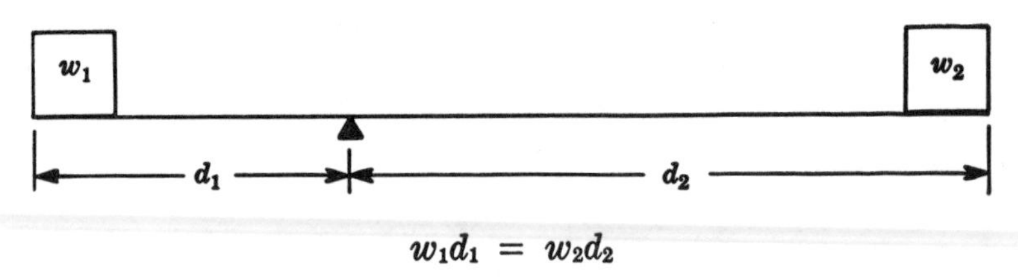

$$w_1 d_1 = w_2 d_2$$

If a 100 pound weight 4 feet from the fulcrum balances a 50 pound weight 8 feet from the fulcrum, then:

$$100 \text{ lbs} \times 4 \text{ feet} = 50 \text{ lbs} \times 8 \text{ feet}$$

$$400 = 400$$

In lever problems it is always wise to draw a sketch to show facts. If you know three of the four facts in the general equation $w_1d_1 = w_2d_2$, you can set up your own equation and solve for the fourth.

EXAMPLE 1. Mary weighed 120 pounds and sat on one end of the teeter-totter 8 feet from the center. Jim weighed 160 pounds and sat on the other side. How far from the fulcrum (balance point) must he sit to balance Mary if the balance point is at the center of the board?

Solution

1. Read the problem and determine the unknown. The question asks how far from the fulcrum Jim sits, so let

$$x = \text{Jim's distance from fulcrum in feet}$$

2. First fact: Mary weighs 120 pounds and sits on one end of the teeter-totter.

3. Second fact: She sits 8 feet from the center.

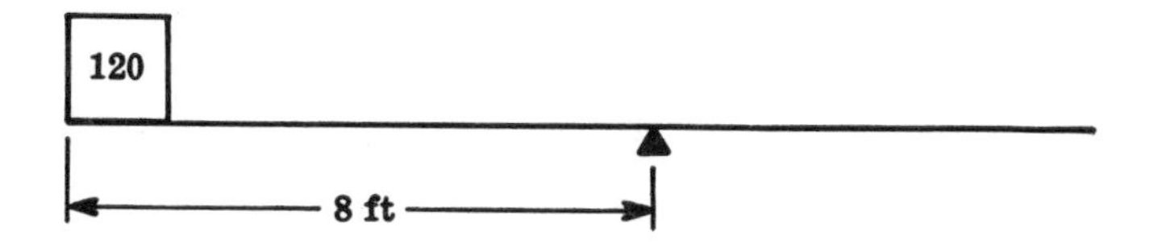

4. Third fact: Jim weighed 160 pounds and sat on the other side. We have already found that we would let x equal Jim's distance.

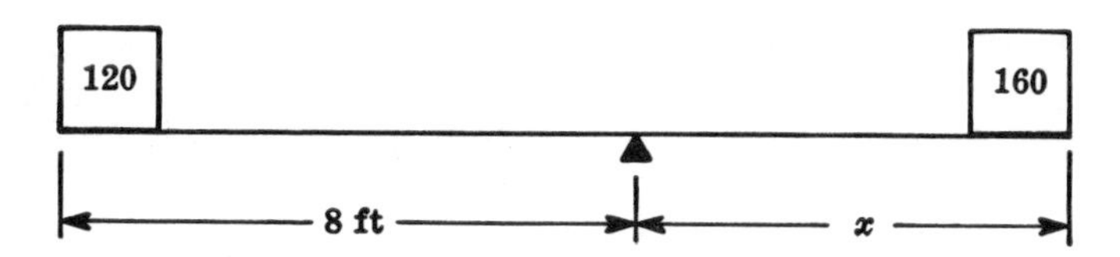

Equation

$$w_1d_1 = w_2d_2$$
$$160x = 120(8)$$
$$160x = 960$$
$$x = 6 \quad \textit{Answer}$$

EXAMPLE 2. Tim was digging up his yard to plant a garden. He hit a large rock and got a crowbar to raise the rock to the surface. The crowbar was 6 feet long and the rock was later found to weigh 80 pounds. Tim used another rock as a fulcrum and exerted a force equal to 16 pounds on the end of the crowbar in order to raise the rock to a point of balance. How far was the fulcrum from the 80-pound rock?

Solution

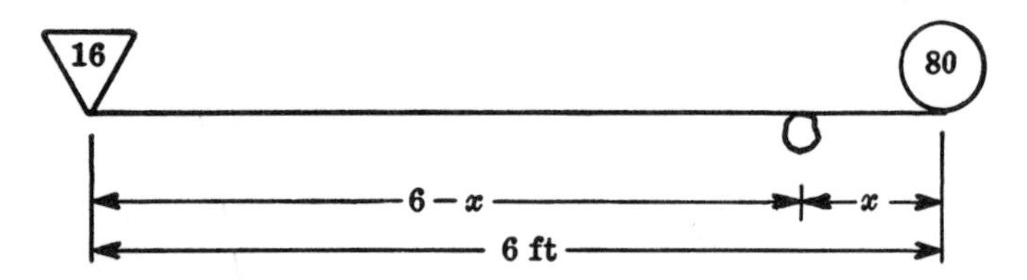

Notice that we again use the principle that the total minus one part equals the second part.

Let x = distance in feet of 80-lb rock from fulcrum

Equation

$$w_1d_1 = w_2d_2$$
$$16(6-x) = 80x$$
$$96 - 16x = 80x$$
$$-96x = -96$$
$$x = 1 \quad \textit{Answer}$$

EXAMPLE 3. A weight of 60 pounds rests on the end of an 8-foot lever and is 3 feet from the fulcrum. What weight must be placed on the other end of the lever to balance the 60-pound weight?

Solution

If the lever is 8 feet long and one end is 3 feet from the fulcrum, the other end must be 5 feet from the fulcrum $(8-3=5)$.

Let x = unknown weight in pounds

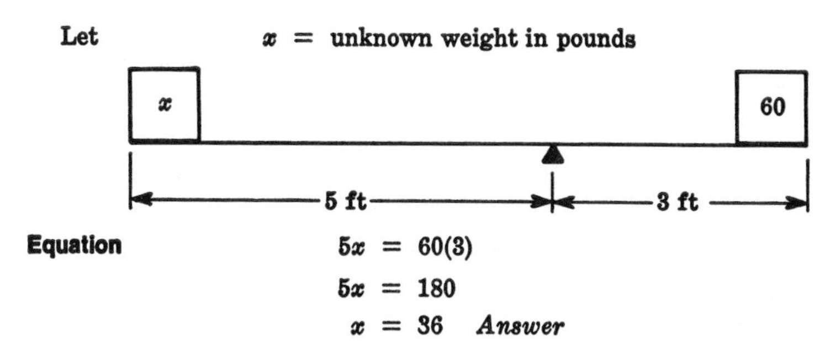

Equation

$$5x = 60(3)$$
$$5x = 180$$
$$x = 36 \quad \textit{Answer}$$

EXAMPLE 4. Four girls decided to use the same teeter-totter. The ones weighing 75 pounds and 50 pounds sat on opposite *ends* of the 12-foot board which had the balance point at the center. The one weighing 60 pounds got on the same side as the one weighing 50 pounds and sat 5 feet from the fulcrum. Where must the fourth girl sit so they balance if *she* weighs 40 pounds?

Solution

Here we have something new. There is more than one weight on the same side. When this happens, you must add the *products* on the same side or

$$w_1d_1 + w_2d_2 = w_3d_3 + w_4d_4$$

Each weight is multiplied by *its* distance from the fulcrum. Let's take it a piece at a time. The ones weighing 75 and 50 pounds sat on opposite ends of a 12-foot board.

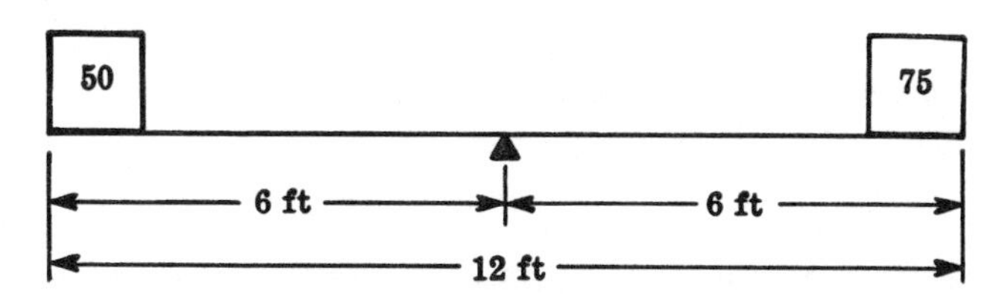

The one weighing 60 pounds got on the same side as the one weighing 50 pounds and sat 5 feet from the fulcrum.

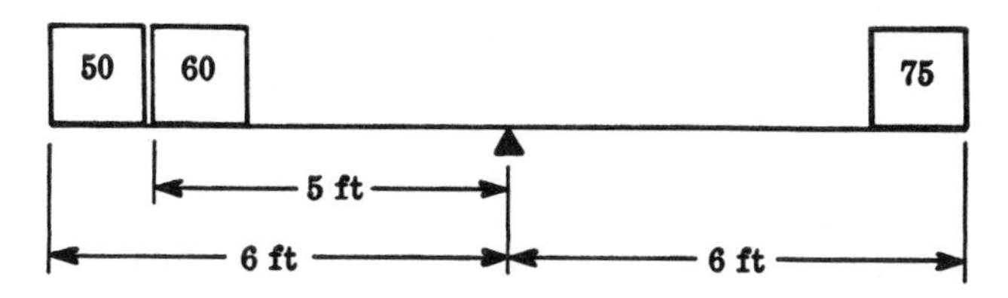

$$w_1d_1 + w_2d_2 = w_3d_3$$
$$50(6) + 60(5) = 75(6)$$

If this equation is true, they balance. But 300 plus 300 is not equal to 450, so the fourth girl must sit on the right side. Her distance will be equal to x because that is the one fact we do not know. So, let

x = distance of 40-lb girl from fulcrum in feet

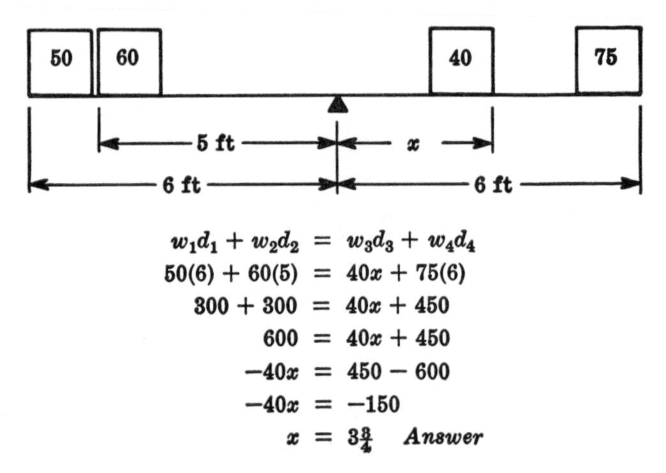

$$w_1d_1 + w_2d_2 = w_3d_3 + w_4d_4$$
$$50(6) + 60(5) = 40x + 75(6)$$
$$300 + 300 = 40x + 450$$
$$600 = 40x + 450$$
$$-40x = 450 - 600$$
$$-40x = -150$$
$$x = 3\tfrac{3}{4} \quad \textit{Answer}$$

SUPPLEMENTARY LEVER PROBLEMS

1. A 200-pound weight rests on one end of a lever, 12 feet from the fulcrum. What weight, resting on the opposite end and 3 feet from the fulcrum, would make a balance?

2. The boys take a 12-foot long board and rest it on a large rock to make a teeter-totter. If the boys sit on opposite ends of the board and weigh 50 pounds and 70 pounds respectively, how far from the fulcrum will the 70-pound boy be if they balance?

3. A window in Mr. Jones' house is stuck. He takes an 8-inch screwdriver to pry open the window. If the screwdriver rests on the sill (fulcrum) 3 inches from the window and Mr. Jones has to exert a force of 10 pounds on the other end to pry open the window, how much force was the window exerting?

4. A 35-pound weight is 2 feet from the fulcrum, and a 75-pound weight on the same side is 10 feet from the fulcrum. If a weight on the other end 6 feet from the fulcrum balances the first two, how much does it weigh?

5. A lever 10 feet long has a 100-pound weight on one end and a 150-pound weight on the other. If the fulcrum is in the center, in what location must an 80-pound weight be placed so that the lever will balance?

6. Shelly and Karie go out to play. Shelly, who weighs 90 pounds, sits on one end of a 14-foot teeter-totter. Its balance point is at the center of the board. Karie, who weighs 120 pounds, climbs on the other end and slides toward the center until they balance. What is Karie's distance *from her end* of the teeter-totter when they balance?

SOLUTIONS TO SUPPLEMENTARY LEVER PROBLEMS

1. Let x = weight in pounds to make a balance

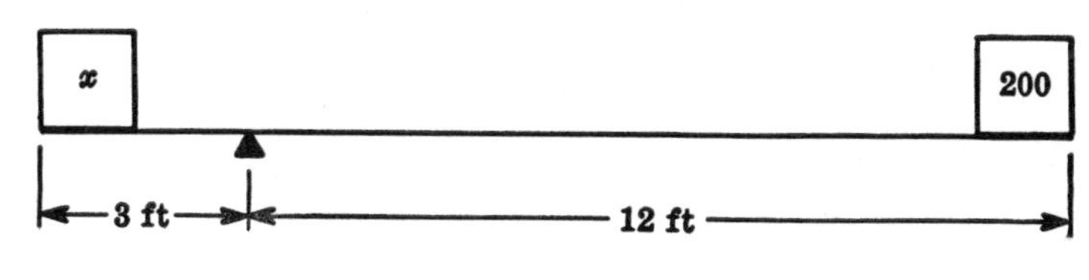

$$3x = 12(200)$$
$$3x = 2400$$
$$x = 800 \quad \textit{Answer}$$

2. Let x = distance in feet of 70-lb boy from fulcrum

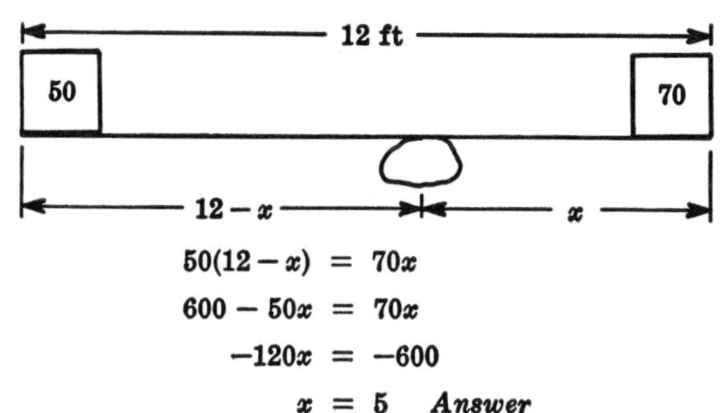

$$50(12 - x) = 70x$$
$$600 - 50x = 70x$$
$$-120x = -600$$
$$x = 5 \quad \textit{Answer}$$

3. Let x = force the window exerts in pounds

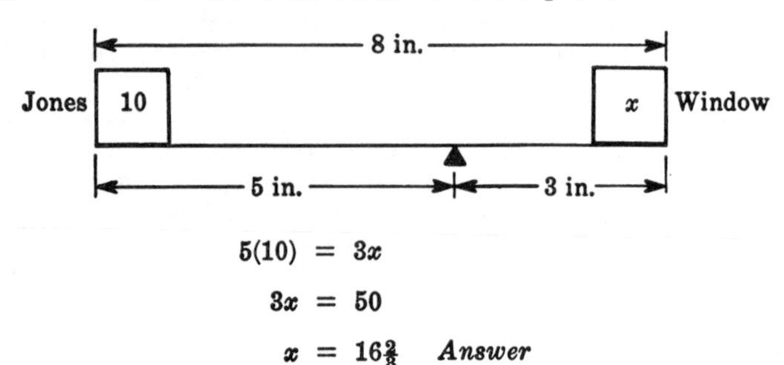

$$5(10) = 3x$$
$$3x = 50$$
$$x = 16\tfrac{2}{3} \quad \textit{Answer}$$

4. Let x = weight in pounds of the weight 6 ft from fulcrum

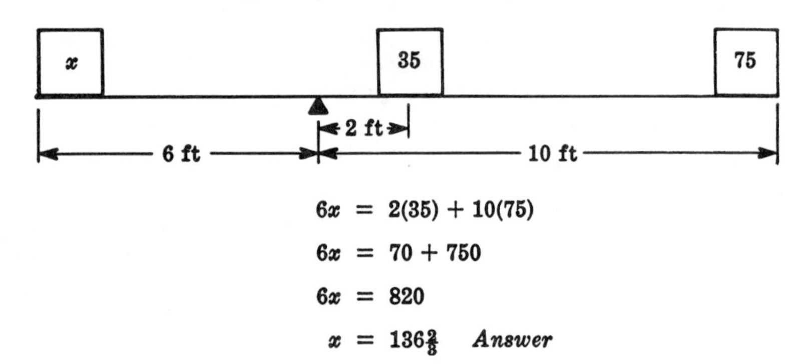

$$6x = 2(35) + 10(75)$$
$$6x = 70 + 750$$
$$6x = 820$$
$$x = 136\tfrac{2}{3} \quad \textit{Answer}$$

5.

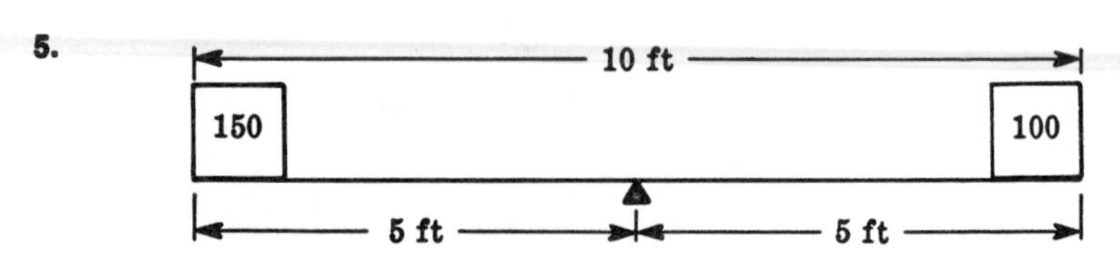

Find products:

$$5(150) = 750 \text{ on left}$$
$$5(100) = 500 \text{ on right}$$

Therefore extra weight must be put on right side to maintain balance.

Let x = distance of 80-lb weight from fulcrum in feet

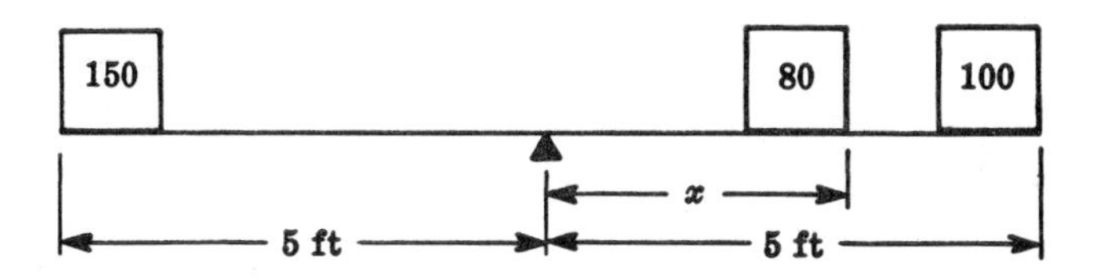

Equation

$$5(150) = 80x + 5(100)$$
$$750 = 80x + 500$$
$$80x + 500 = 750$$
$$80x = 250$$
$$x = 3\tfrac{1}{8} \quad \textit{Answer}$$

6. Let x = Karie's distance from the center in feet

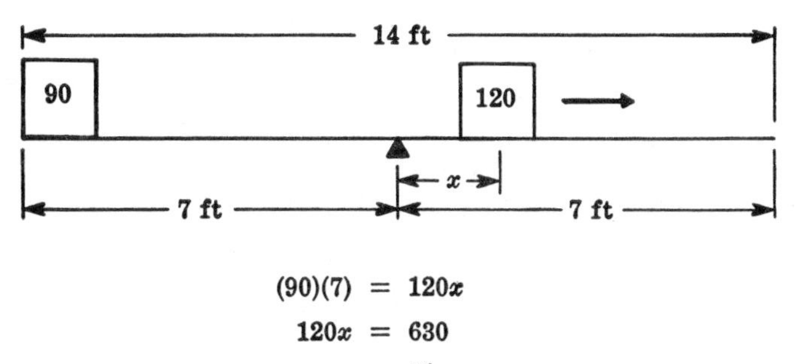

$$(90)(7) = 120x$$
$$120x = 630$$
$$x = 5\tfrac{1}{4}$$
$$7 - 5\tfrac{1}{4} = 1\tfrac{3}{4} \quad \textit{Answer}$$

The question this time asked for the distance of the 120-pound weight (Karie's weight) from the *end*. You must find the distance from the fulcrum first and subtract this distance from the total of 7 feet to find distance from the end of the teeter-totter.

Chapter 7

Finance

There are many types of problems involving money. They should be easy to understand because their principles are familiar to everyone. Let's start with some investment problems. You should know that investments involve three things: *the amount of money invested* (or principal), the *rate of interest* (or percent), and the actual yearly *interest in dollars.* This yearly interest or income is often referred to as the *return* on investment. *Interest, income,* and *return* all mean the same for these problems and any may be used here. (We shall assume simple interest yearly.)

EXAMPLE 1. Mr. McGregor invested \$10,000. Part of it he put in the bank at 5% interest. The remainder he put in bonds which paid a 9% yearly return. How much did he invest in each if his yearly income from the two investments was \$660?

Solution

Let x = amount in dollars invested at 5%

$10{,}000 - x$ = amount in dollars invested at 9%

Now that you have represented the unknowns, using x and $total - x$, and labeling them with their *percent* return, you have to convert the information into *income.* Here is the way you find income (interest): Principal times rate equals interest.

If you put \$1000 in the bank at 5%, your interest would be 0.05(1000). So

$0.05x$ = interest from bank investment

$0.09(10{,}000 - x)$ = interest from bond investment

Equation

$$0.05x + 0.09(10{,}000 - x) = 660$$

In solving an equation containing parentheses and decimals, first clear parentheses and second clear decimals. This helps to avoid careless errors in decimals.

$$0.05x + 900 - 0.09x = 660$$

Multiply by 100 to clear decimals.

$$5x + 90{,}000 - 9x = 66{,}000$$
$$-4x = -24{,}000$$
$$\left.\begin{aligned} x &= 6000 \\ 10{,}000 - x &= 4000 \end{aligned}\right\} \textit{Answers}$$

Check

$$0.05(6000) + 900 - 0.09(6000) = 660$$
$$300 + 900 - 540 = 660$$
$$660 = 660$$

EXAMPLE 2. Mr. Gold invested \$50,000, part at 6% and part at 8%. The annual interest on the 6% investment was \$480 more than that from the 8% investment. How much was invested at each rate?

Solution

Let x = amount in dollars invested at 6%

$50{,}000 - x$ = amount in dollars invested at 8%

$0.06\,x$ = interest on 1st investment

$0.08(50{,}000 - x)$ = interest on 2nd investment

The interest on the 6% investment was \$480 more than the interest on 8%.

Equation

$$0.06\,x = 0.08(50{,}000 - x) + 480$$
$$0.06\,x = 4000 - 0.08\,x + 480$$
$$0.06\,x = 4480 - 0.08\,x$$
$$6x = 448{,}000 - 8x$$
$$14x = 448{,}000$$
$$\left.\begin{aligned} x &= 32{,}000 \\ 50{,}000 - x &= 18{,}000 \end{aligned}\right\} \textit{Answers}$$

Check

$$0.06(32{,}000) = 0.08(50{,}000 - 32{,}000) + 480$$
$$1920 = 4000 - 2560 + 480$$
$$1920 = 1920$$

EXAMPLE 3. A store advertised men's suits on sale at 20% off. The sale price was \$76. What was the original price?

Solution

Let x = original price in dollars

If the sale price is 20% *off* the original price, the sale price is 80% *of* the original price.

Equation

$$0.80\,x = 76$$
$$80x = 7600$$
$$x = 95 \quad \textit{Answer}$$

EXAMPLE 4. Tickets to the school play sold at \$2 each for adults and 75¢ each for children. If there were four times as many adult tickets sold as children's tickets, and the total receipts were \$1750, how many children's tickets were sold?

Solution

Let x = number of children's tickets at 75¢ each

Then $4x$ = number of adult tickets at \$2 each

$0.75\,x$ = total amount of money received from children's tickets in dollars

$2(4x)$ = total amount of money received from adult tickets in dollars

Equation

$$0.75\,x + 2(4x) = 1750$$
$$0.75\,x + 8x = 1750$$
$$75x + 800x = 175{,}000$$
$$875x = 175{,}000$$
$$x = 200 \quad \textit{Answer}$$

EXAMPLE 5. Mr. Herkowitz owns a jewelry store. He marks up all merchandise 150% of cost. If he sold a diamond ring for \$2250, what did he pay the wholesaler for it?

Solution

Let x = amount in dollars he paid the wholesaler

$1.50x$ = markup

Cost plus markup equals selling price

Equation

$$x + 1.50x = 2250$$

Multiply by 10 to clear decimals.

$$10x + 15x = 22{,}500$$

$$25x = 22{,}500$$

$$x = \$900 \quad \textit{Answer}$$

EXAMPLE 6. What amount of money invested at $8\frac{1}{4}\%$ yields \$2475 per year?

Solution

Let x = amount of money invested (in dollars)

$0.0825x$ = interest of $8\frac{1}{4}\%$ per year

Equation

$$0.0825x = 2475$$

Multiply by 10,000 to clear decimals.

$$825x = 24{,}750{,}000$$

$$x = \$30{,}000 \quad \textit{Answer}$$

Check

$$0.0825x = 2{,}475$$

$$0.0825(30{,}000) = 2{,}475$$

$$2{,}475 = 2{,}475$$

EXAMPLE 7. Large supermarkets often price miscellaneous merchandise at a price they know will sell fast and determine what they can pay for it by this selling price. The Hoopla Market bought men's T shirts to sell at \$10 each. If they allowed 40% of the *selling price* for expenses and profit, what would they be willing to pay for the shirts?

Solution

Cost plus markup equals selling price

Let x = cost in dollars they would pay for each shirt

$0.40(10)$ = markup (40% of selling price)

Equation

$$x + 0.40(10) = 10$$

$$x + 4 = 10$$

$$x = \$6 \quad \textit{Answer}$$

Check

$$6 + 0.40(10) = 10$$
$$6 + 4 = 10$$
$$10 = 10$$

Alternate Solution

Let x = cost in dollars of each shirt

Cost + markup = selling price

If the markup is 40% of selling price, the cost is 60% of the selling price.

Cost (60%) + markup (40%) = selling price (100%)

So, cost = 60% of selling price

Equation

$$x = 0.60(10)$$
$$x = \$6 \qquad \textit{Answer}$$

SUPPLEMENTARY FINANCE PROBLEMS

1. Mr. Ritchswitch invested $50,000. Part of it he put in a gold mine stock from which he hoped to receive 20% per year. The rest he invested in bank stock which was paying 6%. If he received $400 more the first year from the bank stock than from the mining stock, how much did he invest in each stock?

2. Mr. McGillicuddy wished to invest a sum of money so that the interest each year would pay for his son's college expenses. If the money was invested at 8% and the college expenses were $10,000 per year, how much should Mr. McGillicuddy invest?

3. The Fleabag had a sale on Indian shirts at which all shirts sold at 15% off. Casey purchased a shirt for $7.65. What was the original selling price? (Assume no tax.)

4. Theodore inherited two different stocks whose yearly income was $2100. The total appraised value of the stocks was $40,000 and one was paying 4% and one 6% per year. What was the value of each stock?

5. A store bought 500 suits, some at $125 each and the rest at $200 each. If the total cost of the suits was $77,500, how many suits were purchased at each price?

6. When Mrs. Oglethorpe sold her house recently, she received \$210,000 for it. This was 40% more than she paid for it 10 years ago. What was the original purchase price?

7. Yi-fei Wang inherited \$20,000 which she invested in stocks and bonds. The stocks returned 6% and the bonds 8%. If the return on the bonds was \$80 less than the return on the stocks, how much did she invest in each?

8. The total of two investments is \$25,000. One amount is invested at 7% and one at 9%. The annual interest from the 7% investment is \$470 more than from the 9% investment. How much is invested at each rate?

9. A taxpayer's state and federal income taxes plus an inheritance tax totaled \$14,270. His California state income tax was \$5780 less than his federal tax. His inheritance tax was \$2750. How much did he pay in state and federal taxes?

10. Mrs. Scott had saved \$6000 which she wished to invest. She put part in a term bank savings account at 8% and part in a regular savings account at $5\frac{1}{2}$%. How much was invested in each account if her total yearly income amounted to \$425?

11. Dr. Williams had \$10,000 invested at 5%. How many dollars would he have to invest at 8% so that his total interest per year would equal 7% of the two investments?

SOLUTIONS TO SUPPLEMENTARY FINANCE PROBLEMS

1. Let

$$x = \text{amount in dollars invested at 20\%}$$
$$50{,}000 - x = \text{amount in dollars invested at 6\% (total minus } x\text{)}$$

$$0.20x = \text{interest on mining stock (smaller income)}$$
$$0.06(50{,}000 - x) = \text{interest on bank stock (larger income)}$$

Equation

Income from bank stock is interest on mining stock plus \$400 more.

$$0.06(50{,}000 - x) = 0.20x + 400$$
$$3000 - 0.06x = 0.20x + 400$$
$$300{,}000 - 6x = 20x + 40{,}000$$
$$-26x = -260{,}000$$
$$\left.\begin{aligned} x &= \$10{,}000 \\ 50{,}000 - x &= \$40{,}000 \end{aligned}\right\} \quad \textit{Answers}$$

Check

$$0.06(50{,}000 - 10{,}000) = 0.20(10{,}000) + 400$$
$$3000 - 600 = 2000 + 400$$
$$2400 = 2400$$

2. Let x = amount in dollars Mac should invest

$0.08x$ = interest per year (assuming simple interest)

Equation

$$0.08x = 10{,}000$$
$$8x = 1{,}000{,}000$$
$$x = \$125{,}000 \text{ invested per year} \qquad \textit{Answer}$$

Check

$$0.08(125{,}000) = 10{,}000$$
$$10{,}000 = 10{,}000$$

3. Let x = original selling price in dollars

$0.15\,x$ = discount

original price − discount = sale price

Equation

$$x - 0.15x = 7.65$$
$$0.85x = 7.65$$
$$85x = 765$$
$$x = \$9 \qquad \textit{Answer}$$

Check

$$9 - 0.15(9) = 7.65$$
$$9 - 1.35 = 7.65$$
$$7.65 = 7.65$$

4. Let x = value in dollars of stock paying 4%

$40{,}000 - x$ = value in dollars of stock paying 6%

$0.04\,x$ = interest on stock paying 4%

$0.06(40{,}000 - x)$ = interest on stock paying 6%

Equation

$$0.04x + 0.06(40{,}000 - x) = 2100$$
$$0.04x + 2400 - 0.06x = 2100$$
$$4x + 240{,}000 - 6x = 210{,}000$$
$$-2x = -30{,}000$$
$$\left.\begin{aligned} x &= \$15{,}000 \\ 40{,}000 - x &= \$25{,}000 \end{aligned}\right\} \textit{Answers}$$

Check

$$0.04(15{,}000) + 0.06(40{,}000 - 15{,}000) = 2{,}100$$
$$600 + 2{,}400 - 900 = 2{,}100$$
$$2{,}100 = 2{,}100$$

5. Let x = number of suits at \$125 each

$500 - x$ = number of suits at \$200 each

total value $125 \times 200(500 - x)$

Equation

$$125x + 200(500 - x) = 77{,}500$$
$$125x + 100{,}000 - 200x = 77{,}500$$
$$-75x = -22{,}500$$
$$\left.\begin{aligned} x &= 300 \text{ suits at \$125 each} \\ 500 - x &= 200 \text{ suits at \$200 each} \end{aligned}\right\} \textit{Answers}$$

Check

$$125(300) + 200(200) = 77{,}500$$
$$77{,}500 = 77{,}500$$

6. Let x = original purchase price

$0.40x$ = increase in price

Equation

$$x + 0.40x = 210{,}000$$
$$1.40x = 210{,}000$$
$$14x = 2{,}100{,}000$$
$$x = \$150{,}000, \text{ original purchase price} \qquad \textit{Answer}$$

Check

$$150{,}000 + 0.4(150{,}000) = 210{,}000$$
$$150{,}000 + 60{,}000 = 210{,}000$$
$$210{,}000 = 210{,}000$$

7. Let

$$x = \text{amount in dollars invested in stock at 6\%}$$
$$20{,}000 - x = \text{amount in dollars invested in bonds at 8\%}$$

$$\mathbf{0.08\,x = \text{interest on stocks}}$$
$$\mathbf{0.06(20{,}000 - x) = \text{interest on bonds}}$$

Equation

Interest on stocks is interest on bonds less $80.

$$0.06x = 0.08(20{,}000 - x) + 80$$
$$0.06x = 1600 - 0.08x + 80$$
$$6x = 160{,}000 - 8x - 8000$$
$$14x = 168{,}000$$
$$\left.\begin{aligned} x &= \$12{,}000 \text{ in stocks} \\ 20{,}000 - x &= \$8000 \text{ in bonds} \end{aligned}\right\} \quad \textit{Answers}$$

Check

$$0.06(12{,}000) = 0.08(20{,}000 - 12{,}000) + 80$$
$$720 = 1600 - 960 + 80$$
$$720 = 720$$

8. Let

$$x = \text{amount invested at 7\%}$$
$$25{,}000 - x = \text{amount invested at 9\%}$$

Equation

$$0.07x = 0.09(25{,}000 - x) + 470$$
$$0.07x = 2250 - 0.09x + 470$$
$$7x = 225{,}000 - 9x + 47{,}000$$
$$16x = 272{,}000$$
$$\left.\begin{aligned} x &= \$17{,}000, \text{ amount invested at 7\%} \\ 25{,}000 - x &= \$8000, \text{ amount invested at 9\%} \end{aligned}\right\} \quad \textit{Answers}$$

Check

$$0.07(17{,}000) = 0.09(25{,}000 - 17{,}000) + 470$$
$$1190 = 720 + 470$$
$$1190 = 1190$$

9. Let x = amount of state tax in dollars

$x + 5780$ = amount of federal tax in dollars

Equation

$$x + x + 5780 + 2750 = 14{,}270$$
$$2x + 8530 = 14{,}270$$
$$2x = 5740$$

$x = \$2870$, amount of state tax
$x + 5780 = \$8650$, amount of federal tax
} *Answers*

Check

$$2870 + 2870 + 5780 + 2750 = 14{,}270$$
$$14{,}270 = 14{,}270$$

10. Let x = amount in dollars invested at 8%

$6000 - x$ = amount in dollars invested at $5\frac{1}{2}\%$

$0.08x$ = interest on 8% investment

$0.055(6000 - x)$ = interest on $5\frac{1}{2}\%$ investment

Equation

Total interest (income) was $425.

$$0.08x + 0.055(6000 - x) = 425$$
$$0.08x + 330 - 0.055x = 425$$

Multiply by 1000 to clear decimals.

$$80x + 330{,}000 - 55x = 425{,}000$$
$$25x = 95{,}000$$

$x = \$3800$
$6000 - x = \$2200$
} *Answers*

Check

$$0.08(3800) + 0.055(6000 - 3800) = 425$$
$$304 + 330 - 209 = 425$$
$$425 = 425$$

11. Let x = amount in dollars at 8%

$$10{,}000 = \text{total amount invested at } 5\%$$
$$x + 10{,}000 = \text{total investment at } 7\%$$

$$0.08\,x = \text{interest on amount at } 8\%$$
$$0.05(10{,}000) = \text{interest on amount at } 5\%$$
$$0.07(x + 10{,}000) = \text{interest on entire investment at } 7\%$$

Equation

Total interest per year equals 7% of entire investment.

$$0.08\,x + 0.05(10{,}000) = 0.07(x + 10{,}000)$$
$$0.08\,x + 500 = 0.07\,x + 700$$

Multiply by 100.

$$8x + 50{,}000 = 7x + 70{,}000$$
$$x = \$20{,}000 \qquad \textit{Answer}$$

Check

$$0.08(20{,}000) + 0.05(10{,}000) = 0.07(20{,}000 + 10{,}000)$$
$$1600 + 500 = 1400 + 700$$
$$2100 = 2100$$

Chapter 8

Work

Now let's take a look at problems involving people or machines doing a job. These are called work problems. Laborers may be digging a ditch, secretaries typing papers, pipes filling a tank. We shall start with basic problems. For example, Mr. Jones can build a garage in 3 days, and Mr. Smith can build the *same* garage in 5 days. How long would it take them to build the garage working together?

There are certain facts about work problems to remember.

1. The jobs in each problem are either the same or equivalent jobs. In the problem above, Smith builds a garage exactly the same as the one Jones builds.

2. In solving a work problem, you need to work with the *same* units of measure, such as minutes, hours, days, etc., within each problem. You can't mix minutes and hours in the same equation, for example.

3. You always need to find the *fractional part* of the job which would be done in one unit of time, such as 1 minute or 1 hour. If a person can do a complete job in 3 days, he can do one-third of it in 1 day.

4. The fractional part of the job one person can do in 1 day plus the fractional part another person can do in 1 day equals the fractional part of the job the two can do together in 1 day. If Mr. Jones builds one-third of the garage in 1 day and Mr. Smith builds one-fifth of it in 1 day, together they can build one-third plus one-fifth of the garage in 1 day.

Solution

We can make a diagram for this type of problem also. Suppose we wanted to know how many days it would take them to build the garage together.

Let x = number of days to build the garage together

	Total time in days	Fractional part of job done in 1 day
Jones	3	$\frac{1}{3}$
Smith	5	$\frac{1}{5}$
Together	x	$\frac{1}{x}$

If you add the fractional parts of the job the two can do separately in 1 day, the answer will be equal to the fractional part of the job the two can do together in 1 day.

Equation

$$\frac{1}{3}+\frac{1}{5} = \frac{1}{x}$$

Multiply by LCD, $15x$, to clear fractions.

$$5x + 3x = 15$$
$$8x = 15$$
$$x = 1\tfrac{7}{8} \quad \textit{Answer}$$

EXAMPLE 1. Bob can dig a ditch in 4 hours. John can dig the same ditch in 3 hours. How long would it take them to dig it together?

Solution

Let x = number of hours to dig the ditch together

	Total time in hours	Fractional part of job done in 1 hour
Bob	4	$\frac{1}{4}$
John	3	$\frac{1}{3}$
Together	x	$\frac{1}{x}$

If Bob takes 4 hours to dig the ditch, he can dig one-fourth of it in 1 hour. John can dig one-third of it in one hour. If it takes them x hours to dig it together, they can dig $1/x$ part of it in 1 hour together. The total of the fractional part each can dig or one-third plus one-fourth equals the fractional part they can dig together in 1 hour or $1/x$.

Equation

$$\frac{1}{3}+\frac{1}{4}=\frac{1}{x}$$

Multiply by LCD, $12x$, to clear fractions.

$$4x + 3x = 12$$

$$7x = 12$$

$$x = 1\tfrac{5}{7} \quad \textit{Answer}$$

EXAMPLE 2. If John can type a paper in 5 hours and together he and Jim can type it in 2 hours, how long would it take Jim to type the same paper alone?

Solution

Let x = number of hours for Jim to type alone

	Total time in hours	Fractional part of job done in 1 hour
John	5	$\frac{1}{5}$
Jim	x	$\frac{1}{x}$
Together	2	$\frac{1}{2}$

Equation

$$\frac{1}{5}+\frac{1}{x}=\frac{1}{2}$$

$$2x + 10 = 5x$$

$$-3x = -10$$

$$x = 3\tfrac{1}{3} \quad \textit{Answer}$$

From these basic problems we can go to more complicated ones, but we will always work with fractional parts of a job done in 1 day, 1 hour, etc. The complete job is equal to the sum of the fractional parts or is equal to "1." If A does one-quarter, B does one-quarter, and C does one-half, the whole job is done, and $\frac{1}{4}+\frac{1}{4}+\frac{1}{2}=1$.

EXAMPLE 3. Two women paint a barn. Barbara can paint it alone in 5 days, Sara in 8 days. They start to paint it together, but after 2 days Sara gets bored and Barbara finishes alone. How long does it take her to finish?

Solution

All work problems need diagrams. In the first column, show the *total time* for each person to do the job. Then in the next column show the fractional part of the job each person can do in *one* unit of time.

Let x = number of days for Barbara to finish barn

	Total time in days	Fractional part of job done in 1 day
Barbara	5	$\frac{1}{5}$
Sara	8	$\frac{1}{8}$

Next, for the third column multiply the time each person actually works by the column marked *fractional part of job done in 1 day*. The resulting fraction in the third column is the part of the job each person does in the time worked.

	Total time in days	Fractional part of job done in 1 day	Fractional part each does in 2 days	Fractional part Barbara does in x days
Barbara	5	$\frac{1}{5}$	$\frac{2}{5}$	$\frac{x}{5}$
Sara	8	$\frac{1}{8}$	$\frac{2}{8}$	....

Barbara does 2/5, Sara 2/8, and Barbara $x/5$ of the job. Add each part and you get one complete job, equal to 1.

Equation

$$\frac{2}{5}+\frac{2}{8}+\frac{x}{5} = 1$$

Multiply by LCD, 40, to clear fractions.

$$16 + 10 + 8x = 40$$
$$8x + 26 = 40$$
$$8x = 14$$
$$x = 1\tfrac{3}{4} \quad \textit{Answer}$$

EXAMPLE 4. A swimming pool can be filled by an inlet pipe in 10 hours and emptied by an outlet pipe in 12 hours. One day the pool is empty and the owner opens the inlet pipe to fill the pool. But he forgets to close the outlet. With both pipes open, how long will it take to fill the pool?

Solution

This time we have two pipes whose flows work against each other. The filling takes place faster than the emptying. So in any given time, the part of the job done is equal to the amount filled minus the amount emptied.

Let

x = number of hours to fill pool with both pipes open

	Total time in hours	Fractional part filled or emptied in 1 hour	Fractional part each pipe does in x hours
Inlet pipe	10	$\frac{1}{10}$	$\frac{x}{10}$
Outlet pipe	12	$\frac{1}{12}$	$\frac{x}{12}$

Equation

Amount filled minus amount emptied equals whole job.

$$\frac{x}{10}-\frac{x}{12} = 1$$

Multiply by LCD, 60, to clear fractions.

$$6x - 5x = 60$$
$$x = 60 \quad \textit{Answer}$$

EXAMPLE 5. Mary, Sue, and Bill work at a motel. It would take Mary 10 hours (if she worked alone), Sue 8 hours, and Bill 12 hours to clean the whole motel. One day Mary came to work early and she had cleaned for 2 hours when Sue and Bill arrived and all three finished the job. How long did they take to finish?

Solution

Let x = number of hours they took to finish

	Total time in hours	Fractional part of job done in 1 hour	Number of hours worked 2 hours	x hours
Mary	10	$\frac{1}{10}$	$\frac{2}{10}$	$\frac{x}{10}$
Sue	8	$\frac{1}{8}$	....	$\frac{x}{8}$
Bill	12	$\frac{1}{12}$	....	$\frac{x}{12}$

Equation

Fractional parts of job each does added equal total job or 1.

$$\frac{2}{10}+\frac{x}{10}+\frac{x}{8}+\frac{x}{12} = 1$$

Multiply by LCD, 120.

$$24 + 12x + 15x + 10x = 120$$
$$37x = 96$$
$$x = 2\tfrac{22}{37} \quad \textit{Answer}$$

SUPPLEMENTARY WORK PROBLEMS

1. Tom, Dick, and Harry decided to fence a vacant lot adjoining their properties. If it would take Tom 4 days to build the fence, Dick 3 days, and Harry 6 days, how long would it take them working together?

2. John and Charles are linemen. John can string 10 miles of line in 3 days while together they can string it in 1 day. How long would it take Charles alone to string the same line?

3. A firm advertises for workers to address envelopes. Priscilla says she will work 100 hours. Herb will work for 80 hours. If each can address 10,000 envelopes in the time they work, how long would it take them to address 10,000 envelopes if they work together?

4. Mrs. Hensen cataloged a group of incoming library books in 8 hours. Mr. Taylor can catalog the same number of books in 4 hours. Mrs. Phillips can catalog them in 6 hours. How long would it take if they all worked together?

5. One machine can complete a job in 10 minutes. If the same job is done by this machine and an older machine working together, the job can be completed in 6 minutes. How long would it take the older machine to do the job alone?

6. A pond is being drained by a pump. After 3 hours, the pond is half empty. A second pump is put into operation and together the two pumps finish emptying the pond in half an hour. How long would it take the second pump to drain the pond if it had to do the same job alone?

7. Jim and his dad are bricklayers. Jim can lay bricks for a fireplace and chimney in 5 days. With his father's help he can build it in 2 days. How long would it take his father to build it alone?

8. Jones can paint a car in 8 hours. Smith can paint the same car in 6 hours. They start to paint the car together. After 2 hours, Jones leaves for lunch and Smith finishes painting the car alone. How long does it take Smith to finish?

9. Bill, Bob, and Barry are hired to paint signs. In 8 hours Bill can paint 1 sign, Bob can paint 2 signs, and Barry can paint $1\frac{1}{8}$ signs. They all come to work the first day, but Barry doesn't like the job and quits after 3 hours. Bob works half an hour longer than Barry and quits. How long will it take Bill to finish *2 signs* they were supposed to paint?

10. Keith has to grind the valves on his Mercedes. He estimates it will take him 8 hours. His friends Joe and Russ have done the same job in 10 and 12 hours, respectively. Keith had worked for 2 hours when Joe came along and helped him for 2 hours. Joe worked another half an hour alone and then Russ came along and together they finished the job. How long did it take them to finish?

11. Mr. Caleb hires Jay and Debbie to prune his grape vineyard. Jay can prune the vineyard in 50 hours and Debbie can prune it in 40 hours. How many hours will it take them working together?

12. A high school hiking club held a car wash to raise money for equipment. Rudy, Cheryl, Tom, and Pat volunteered to help. Rudy could wash a car in 10 minutes, Cheryl in 12, Tom in 8, and Pat in 15. They all started on the first car, but after 2 minutes another car came in and Pat and Tom went to work on it. One minute later Rudy quit to take care of another customer. How long did it take Cheryl to finish the first car alone?

SOLUTIONS TO SUPPLEMENTARY WORK PROBLEMS

1. Let x = number of days to build fence together

	Total time in days	Fractional part of job done in 1 day
Tom	4	$\frac{1}{4}$
Dick	3	$\frac{1}{3}$
Harry	6	$\frac{1}{6}$
Together	x	$\frac{1}{x}$

Equation

The fractional parts each person does in 1 day equal fractional part done together in 1 day.

$$\frac{1}{4} + \frac{1}{3} + \frac{1}{6} = \frac{1}{x}$$

$$3x + 4x + 2x = 12$$

$$9x = 12$$

$$x = 1\tfrac{1}{3} \quad \textit{Answer}$$

2. Let

x = number of days for Charles to string line alone

	Total time in days	Fractional part of job done in 1 day
John	3	$\frac{1}{3}$
Charles	x	$\frac{1}{x}$
Together	1	$\frac{1}{1}$

Equation

$$\frac{1}{3} + \frac{1}{x} = \frac{1}{1}$$

$$x + 3 = 3x$$

$$-2x = -3$$

$$x = 1\tfrac{1}{2} \quad \textit{Answer}$$

3. Let

x = number of hours to address envelopes together

	Total time in hours	Fractional part of job done in 1 hour
Priscilla	100	$\frac{1}{100}$
Herb	80	$\frac{1}{80}$
Together	x	$\frac{1}{x}$

Equation

$$\frac{1}{100} + \frac{1}{80} = \frac{1}{x}$$

Multiply by LCD, $400x$.

$$4x + 5x = 400$$

$$9x = 400$$

$$x = 44\tfrac{4}{9} \quad \textit{Answer}$$

4. Let x = number of hours for all three to catalog the books working together

	Total time in hours	Fractional part of job done in 1 hour
Hensen	8	$\frac{1}{8}$
Taylor	4	$\frac{1}{4}$
Phillips	6	$\frac{1}{6}$
Together	x	$\frac{1}{x}$

The total of fractional parts of job done by each in one hour equals fractional part done by all of them working together.

Equation

$$\frac{1}{8}+\frac{1}{4}+\frac{1}{6} = \frac{1}{x}$$

Multiply by LCD, $24x$, to clear fractions.

$$3x + 6x + 4x = 24$$
$$13x = 24$$
$$x = 1\tfrac{11}{13} \quad \textit{Answer}$$

5. Let

x = number of minutes for older machine to do job

	Total time in minutes	Fractional part of job done in 1 minute
First machine	10	$\frac{1}{10}$
Older machine	x	$\frac{1}{x}$
Together	6	$\frac{1}{6}$

Equation

$$\frac{1}{10}+\frac{1}{x} = \frac{1}{6}$$

Multiply by LCD, $30x$, to clear fractions.

$$3x + 30 = 5x$$
$$-2x = -30$$
$$x = 15 \quad \textit{Answer}$$

6. If the first pump drains half the pond in 3 hours, it could drain the pond completely in 6 hours. If the two pumps drain the last half in half an hour, together they could drain the whole pond in 1 hour.

Let

x = number of hours it would take second pump to empty pond

	Total time in hours	Fractional part of job done in 1 hour
First pump	6	$\frac{1}{6}$
Second pump	x	$\frac{1}{x}$
Together	1	$\frac{1}{1}$

Equation

$$\frac{1}{6} + \frac{1}{x} = \frac{1}{1}$$

Multiply by LCD, $6x$, to clear fractions.

$$x + 6 = 6x$$
$$-5x = -6$$
$$x = 1\tfrac{1}{5} \quad \textit{Answer}$$

7. Let x = number of days it would take father alone

	Total time in days	Fractional part of job done in 1 day
Jim	5	$\frac{1}{5}$
Father	x	$\frac{1}{x}$
Together	2	$\frac{1}{2}$

Equation

Fractional part Jim can do plus fractional part father can do equal fractional part they can do together.

$$\frac{1}{5} + \frac{1}{x} = \frac{1}{2}$$

Multiply by LCD, $10x$, to clear fractions.

$$2x + 10 = 5x$$
$$-3x = -10$$
$$x = 3\tfrac{1}{3} \quad \textit{Answer}$$

8. Let x = number of hours it takes Smith to finish

	Total time to do job in hours	Fractional part of job done in 1 hour	Fractional part each does in: 2 hours	x hours
Jones	8	$\frac{1}{8}$	$2\left(\frac{1}{8}\right)$	....
Smith	6	$\frac{1}{6}$	$2\left(\frac{1}{6}\right)$	$x\left(\frac{1}{6}\right)$

Each man worked for 2 hours. In that time each did two times the fractional part he could do in 1 hour or $2(\frac{1}{8})$ and $2(\frac{1}{6})$. Mr. Jones left and *only* Mr. Smith worked for x hours so he did $x(\frac{1}{6})$ of the job. When the job is completed, all the fractional parts of the job done add up to 1.

Equation

$$2\left(\frac{1}{8}\right) + 2\left(\frac{1}{6}\right) + x\left(\frac{1}{6}\right) = 1$$
$$\frac{2}{8} + \frac{2}{6} + \frac{x}{6} = 1$$

Reduce fractions.

$$\frac{1}{4} + \frac{1}{3} + \frac{x}{6} = 1$$

Multiply by LCD, 12.

$$3 + 4 + 2x = 12$$
$$2x + 7 = 12$$
$$2x = 5$$
$$x = 2\tfrac{1}{2} \quad \textit{Answer}$$

9. Let x = number of hours it takes Bill to finish 2 signs

The total job in this problem is painting two signs. So the total time for each person is the time it takes him to paint two signs.

Bill can paint 1 sign in 8 hours or 2 signs in 16 hours.

Bob can paint 2 signs in 8 hours.

Barry can paint $1\frac{1}{3}$ signs in 8 hours.

$$8 \div 1\frac{1}{3} = 8 \times \frac{3}{4} = 6 \text{ hours to paint 1 sign}$$

Barry can paint 2 signs in 12 hours.

	Time to paint 2 signs in hours	Fractional part of job done in 1 hour	Fractional part each does in: 3 hours	$\frac{1}{2}$ hour	x hours
Bill	16	$\frac{1}{16}$	$3\left(\frac{1}{16}\right)$	$\frac{1}{2}\left(\frac{1}{16}\right)$	$x\left(\frac{1}{16}\right)$
Bob	8	$\frac{1}{8}$	$3\left(\frac{1}{8}\right)$	$\frac{1}{2}\left(\frac{1}{8}\right)$	....
Barry	12	$\frac{1}{12}$	$3\left(\frac{1}{12}\right)$	....	....

Equation

The total of all fractional parts of job done equals 1.

$$\frac{3}{16} + \frac{3}{8} + \frac{3}{12} + \frac{1}{32} + \frac{1}{16} + \frac{x}{16} = 1$$

Multiply by LCD, 96.

$$18 + 36 + 24 + 3 + 6 + 6x = 96$$
$$6x = 9$$
$$x = 1\tfrac{1}{2} \quad \textit{Answer}$$

10. Let x = number of hours Joe and Russ took to finish the job

	Total time in hours	Fractional part of job done in 1 hour	Fractional part each person does in: 2 hours	2 hours	$\frac{1}{2}$ hour	x hours
Keith	8	$\frac{1}{8}$	$2\left(\frac{1}{8}\right)$	$2\left(\frac{1}{8}\right)$	....	....
Joe	10	$\frac{1}{10}$	....	$2\left(\frac{1}{10}\right)$	$\frac{1}{2}\left(\frac{1}{10}\right)$	$x\left(\frac{1}{10}\right)$
Russ	12	$\frac{1}{12}$	....	....	....	$x\left(\frac{1}{12}\right)$

Equation

The total of all fractional parts of job equals 1.

$$2\left(\frac{1}{8}\right) + 2\left(\frac{1}{8}\right) + 2\left(\frac{1}{10}\right) + \frac{1}{2}\left(\frac{1}{10}\right) + x\left(\frac{1}{10}\right) + x\left(\frac{1}{12}\right) = 1$$

$$\frac{1}{4} + \frac{1}{4} + \frac{1}{5} + \frac{1}{20} + \frac{x}{10} + \frac{x}{12} = 1$$

Multiply by LCD, 60, to clear fractions.

$$15 + 15 + 12 + 3 + 6x + 5x = 60$$

$$11x = 15$$

$$x = 1\tfrac{4}{11} \quad \textit{Answer}$$

11. Let x = number of vines Jay and Debbie can prune together in one hour

	Total time to do job in hours	Fractional part of job done in 1 hour
Jay	50	$\frac{1}{50}$
Debbie	40	$\frac{1}{40}$
Together	x	$\frac{1}{x}$

Equation

The fractional parts of job each can do in one hour equal the fractional part they can do together.

$$\frac{1}{50} + \frac{1}{40} = \frac{1}{x}$$

Multiply by LCD, $200x$, to clear fractions.

$$4x + 5x = 200$$

$$9x = 200$$

$$x = 22\tfrac{2}{9} \quad \textit{Answer}$$

12. Let

x = number of minutes Cheryl took to finish 1st car

	Total time in minutes	Fractional part of job done in 1 minute	Fractional part each does in: 2 minutes	1 minute	x minutes
Rudy	10	$\frac{1}{10}$	$2\left(\frac{1}{10}\right)$	$\frac{1}{10}$	....
Cheryl	12	$\frac{1}{12}$	$2\left(\frac{1}{12}\right)$	$\frac{1}{12}$	$x\left(\frac{1}{12}\right)$
Tom	8	$\frac{1}{8}$	$2\left(\frac{1}{8}\right)$	....	....
Pat	15	$\frac{1}{15}$	$2\left(\frac{1}{15}\right)$	....	....

Sum of fractional parts of job done equals 1.

$$2\left(\frac{1}{10}\right) + 2\left(\frac{1}{12}\right) + 2\left(\frac{1}{8}\right) + 2\left(\frac{1}{15}\right) + \frac{1}{10} + \frac{1}{12} + \frac{x}{12} = 1$$

$$\frac{1}{5} + \frac{1}{6} + \frac{1}{4} + \frac{1}{15} + \frac{1}{10} + \frac{1}{12} + \frac{x}{12} = 1$$

Multiply by LCD, 60, to clear fractions.

$$12 + 10 + 15 + 8 + 6 + 5 + 5x = 60$$

$$5x = 4$$

$$x = \tfrac{4}{5} \quad \textit{Answer}$$

Chapter 9

Plane Geometric Figures

Here are a few facts to remember about plane geometric figures.

1. The perimeter of a rectangle equals two widths plus two lengths.
2. The sum of the three angles of any triangle equals 180°.
3. A square has four equal sides.
4. The area of a rectangle equals length times width.

EXAMPLE 1. The length of a rectangle is equal to twice the width. The perimeter is 138 feet. What are the dimensions?

Solution

Let x = width of rectangle in feet (smaller)

$2x$ = length of rectangle in feet

Here we can again make a sketch.

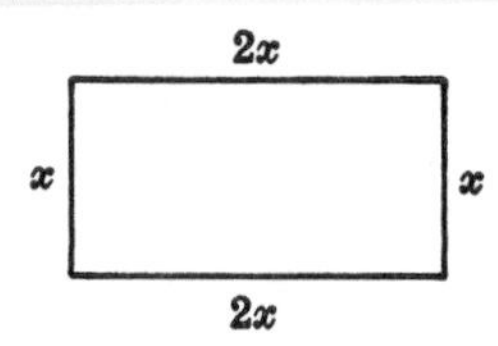

Equation

The perimeter equals two lengths plus two widths.

$$2x + 2x + x + x = 138 \quad \text{or} \quad 2(2x) + 2(x) = 138$$

$$6x = 138$$

$$\left.\begin{aligned} x &= 23 \\ 2x &= 46 \end{aligned}\right\} \textit{Answers}$$

EXAMPLE 2. A rectangle has a length which is 4 feet less than three times the width. The perimeter is 224 feet. What are the dimensions?

Solution

Let

x = width of rectangle in feet

$3x - 4$ = length of rectangle in feet (4 feet less than 3 times width)

Sketch:

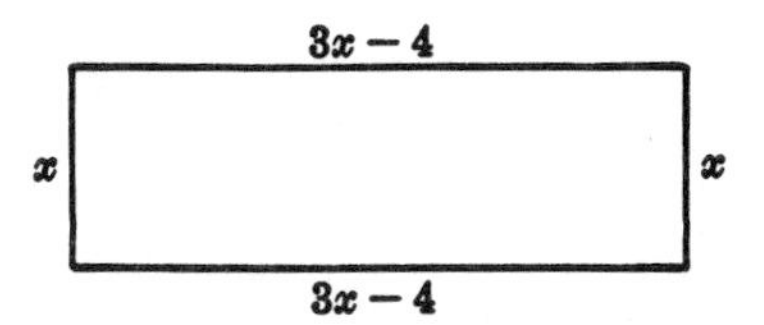

Equation

Perimeter equals two lengths plus two widths.

$$2(x) + 2(3x - 4) = 224$$
$$2x + 6x - 8 = 224$$
$$8x = 232$$
$$\left.\begin{aligned} x &= 29 \\ 3x - 4 &= 83 \end{aligned}\right\} \quad \textit{Answers}$$

EXAMPLE 3. The first angle of a triangle is twice the second and the third is 5 degrees larger than the first. Find the three angles.

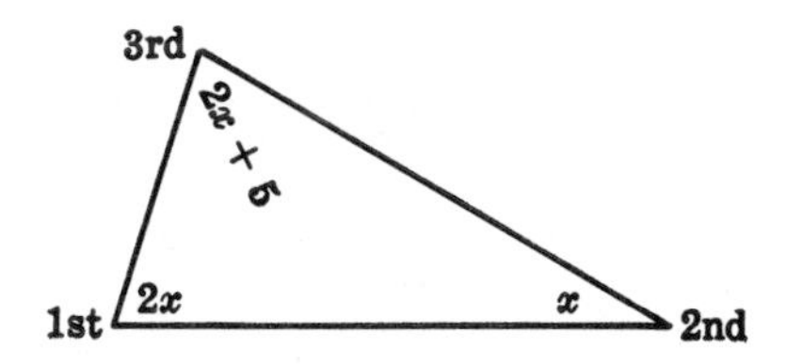

Solution

Let x = number of degrees in second angle (smallest)

$2x$ = number of degrees in first angle

$2x + 5$ = number of degrees in third angle

Equation

Sum of three angles of a triangle equals 180 degrees.

$$x + 2x + 2x + 5 = 180$$
$$5x = 175$$
$$\left.\begin{aligned} x &= 35 \\ 2x &= 70 \\ 2x + 5 &= 75 \end{aligned}\right\} \quad \textit{Answers}$$

EXAMPLE 4. The length of one rectangle is two times the width. If the length is decreased by 5 feet and the width is increased by 5 feet, the area is increased by 75 square feet. Find the dimensions of the original rectangle.

Solution

Here we have two rectangles. Let's draw them.

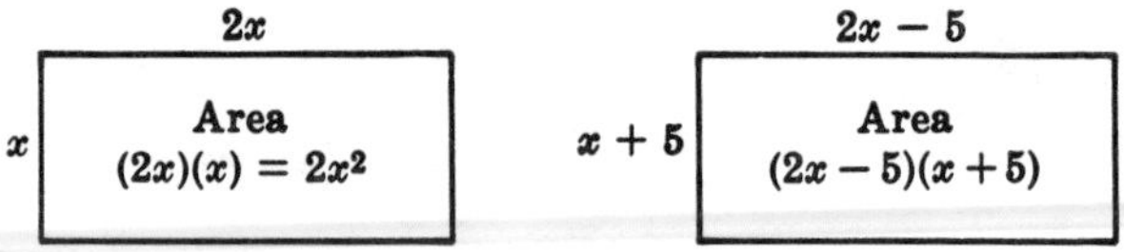

	$2x$		$2x - 5$
x	Area $(2x)(x) = 2x^2$	$x + 5$	Area $(2x-5)(x+5)$

Let x = width of first rectangle in feet

$2x$ = length of first rectangle in feet

Equation

Area of first rectangle plus 75 equals area of second rectangle.

$$2x^2 + 75 = (2x-5)(x+5)$$
$$2x^2 + 75 = 2x^2 + 5x - 25$$

The $2x^2$ terms cancel when all "x's" are transposed to the left.

$$-5x = -100$$
$$\left.\begin{aligned} x &= 20 \\ 2x &= 40 \end{aligned}\right\} \quad \textit{Answers}$$

EXAMPLE 5. The first side of a triangle is 2 inches less than twice the second side. The third side is 10 inches longer than the second side. If the perimeter is 12 feet, find the length of each side.

Solution

Notice that the first and third sides are longer than the second. So we let x equal the second side. Also, note that the perimeter is given in *feet* while the sides are given in *inches*. The perimeter must be changed to inches, and 12 feet equals 144 inches.

Let
$$x = \text{second side in inches}$$
$$2x - 2 = \text{first side in inches}$$
$$x + 10 = \text{third side in inches}$$

Sketch:

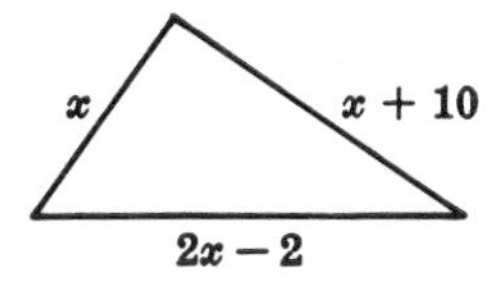

Equation

$$(x) + (2x - 2) + (x + 10) = 144$$
$$4x + 8 = 144$$
$$4x = 136$$
$$\left.\begin{aligned} x &= 34 \\ 2x - 2 &= 66 \\ x + 10 &= 44 \end{aligned}\right\} \quad \textit{Answers}$$

SUPPLEMENTARY PLANE GEOMETRIC FIGURE PROBLEMS

1. The length of a rectangle is 5 feet more than twice the width. The perimeter is 28 feet. Find the length and width.

2. The second angle of a triangle is 20 degrees greater than the first angle. The third is twice the second. Find the three angles.

3. A farmer wishes to fence a rectangular area behind his barn. The barn forms one end of the rectangle and the length of the rectangle is three times the width. How many linear feet of fence must he buy if the perimeter of the rectangle is 320 feet?

4. The first side of a triangle is twice the second, and the third is 20 feet less than three times the second. The perimeter is 106 feet. Find the three sides.

5. The length of a room is 8 feet more than twice the width. If it takes 124 feet of molding to go around the perimeter of the room, what are its dimensions?

6. The perimeter of a triangle is 42 yards. The first side is 5 yards less than the second, and the third is 2 yards less than the first. What is the length of each side?

7. One side of a rectangle is five times as long as the other side. If the perimeter is 72 meters, what is the length of the shorter side?

8. A rectangular box contains 336 cubic inches. It is 12 inches long and 7 inches wide. How deep is it? (Volume equals length times width times depth.)

9. The length of a rectangle is 8 feet more than the width. If the width is increased by 4 feet and the length is decreased by 5 feet, the area remains the same. Find the dimensions of the original rectangle.

SOLUTIONS TO SUPPLEMENTARY PLANE GEOMETRIC FIGURE PROBLEMS

1. Let

$$x = \text{width in feet}$$
$$2x + 5 = \text{length in feet}$$

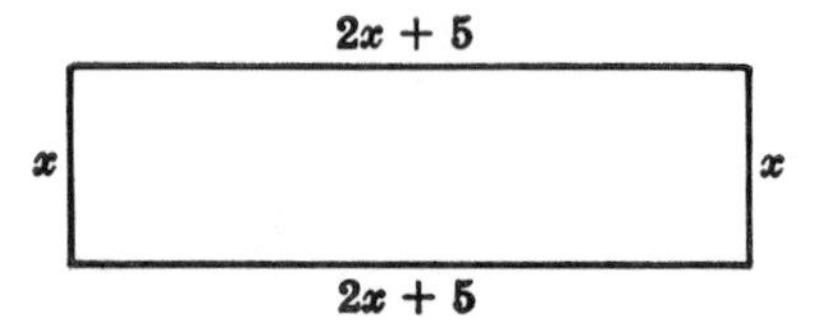

Equation

Perimeter equals two lengths plus two widths.

$$\begin{aligned} 28 &= 2(x) + 2(2x + 5) \\ 28 &= 2x + 4x + 10 \\ 28 &= 6x + 10 \\ -6x &= -18 \\ x &= 3 \\ 2x + 5 &= 11 \end{aligned} \quad \Bigg\} \; \textit{Answers}$$

Check

$$\begin{aligned} 2(3) + 2(11) &= 28 \\ 6 + 22 &= 28 \end{aligned}$$

2. The first angle is smallest.

Let

$$\begin{aligned} x &= \text{number of degrees in first angle} \\ x + 20 &= \text{number of degrees in second angle} \\ 2(x + 20) &= \text{number of degrees in third angle} \end{aligned}$$

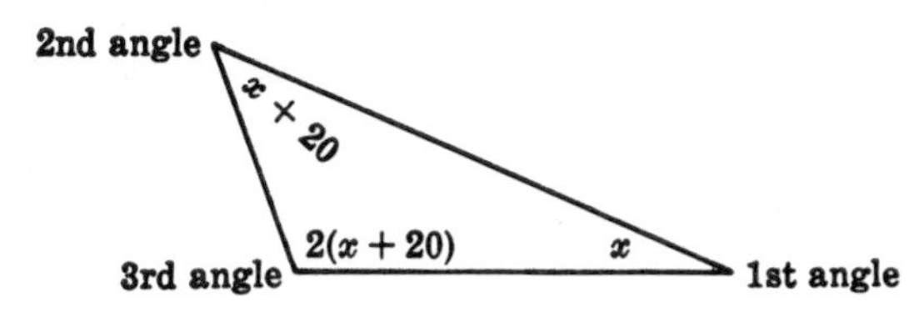

Equation

$$\begin{aligned} x + (x + 20) + 2(x + 20) &= 180 \\ 4x + 60 &= 180 \\ 4x &= 120 \\ x &= 30 \\ x + 20 &= 50 \\ 2(x + 20) &= 100 \end{aligned} \quad \Bigg\} \; \textit{Answers}$$

Check

$$\begin{aligned} 30 + 50 + 100 &= 180 \\ 180 &= 180 \end{aligned}$$

3. Let

$$\begin{aligned} x &= \text{width of fenced area in feet} \\ 3x &= \text{length of fenced area in feet} \end{aligned}$$

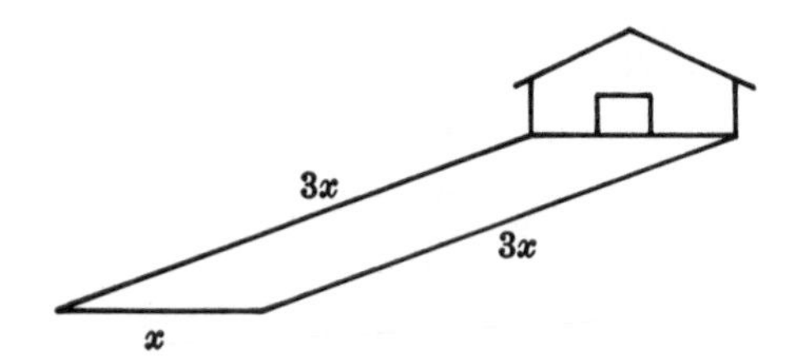

Equation

$$2(3x) + 2(x) = 320$$
$$8x = 320$$
$$x = 40$$
$$3x = 120$$

The barn closes one end of the area. Therefore the farmer needs 2 lengths plus 1 width of fencing or

$$120 + 120 + 40 = 280 \text{ feet}$$ *Answer*

Check

$$40 + 40 + 120 + 120 = 320$$
$$320 = 320$$

4. The second side is smallest.

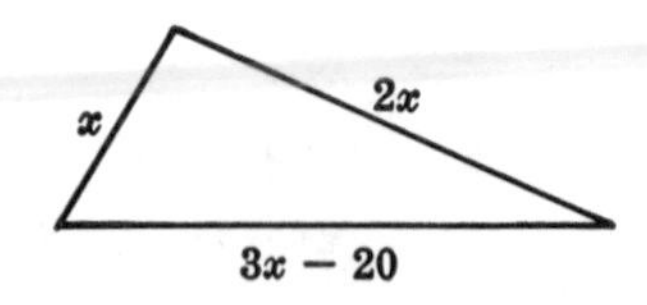

Let

x = second side in feet
$2x$ = first side in feet
$3x - 20$ = third side in feet

Equation

$$x + (2x) + (3x - 20) = 106$$
$$6x = 126$$
$$\left.\begin{aligned} x &= 21 \\ 2x &= 42 \\ 3x - 20 &= 43 \end{aligned}\right\} \textit{Answers}$$

Check

$$21 + 42 + 43 = 106$$
$$106 = 106$$

5. The width is smaller.

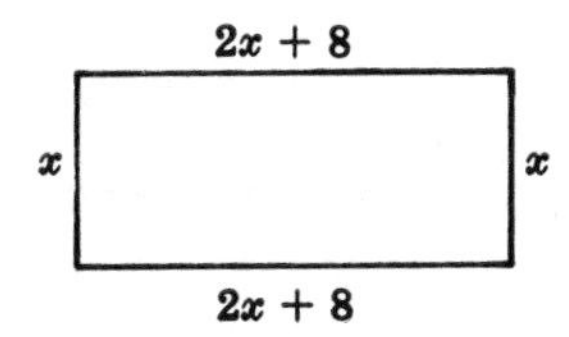

Let

x = width in feet

$2x + 8$ = length in feet (Eight feet more than twice the width equals the length.)

Equation

Perimeter equals two lengths plus two widths.

$$2(x) + 2(2x + 8) = 124$$
$$2x + 4x + 16 = 124$$
$$6x + 16 = 124$$
$$6x = 108$$
$$\left.\begin{aligned} x &= 18 \\ 2x + 8 &= 44 \end{aligned}\right\} \textit{Answers}$$

Check

$$2(18) + 2(44) = 124$$
$$36 + 88 = 124$$
$$124 = 124$$

6. The third side is smallest.

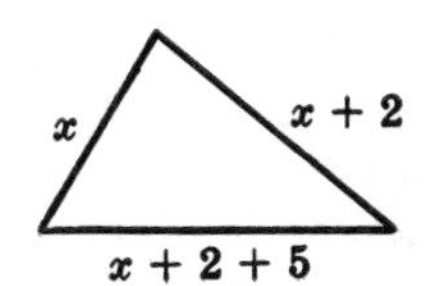

Let

$$\begin{aligned} x &= \text{third side in yards} \\ x + 2 &= \text{first side in yards} \\ (x + 2) + 5 &= \text{second side in yards} \end{aligned}$$

Equation

The sum of 3 sides equals perimeter.

$$\begin{aligned} x + (x + 2) + (x + 2) + 5 &= 42 \\ 3x + 9 &= 42 \\ 3x &= 33 \end{aligned}$$

$$\left.\begin{aligned} x &= 11 \\ x + 2 &= 13 \\ x + 2 + 5 &= 18 \end{aligned}\right\} \quad \textit{Answers}$$

Check

$$\begin{aligned} 11 + 13 + 18 &= 42 \\ 42 &= 42 \end{aligned}$$

7. Let

$$\begin{aligned} x &= \text{width in feet} \\ 5x &= \text{length in feet} \end{aligned}$$

$5x$

x

Equation

$$\begin{aligned} 2(5x) + 2(x) &= 72 \\ 12x &= 72 \end{aligned}$$

$$\left.\begin{aligned} x &= 6 \\ 5x &= 30 \end{aligned}\right\} \quad \textit{Answers}$$

Check

$$\begin{aligned} 2(6) + 2(30) &= 72 \\ 72 &= 72 \end{aligned}$$

8. Let

$$x = \text{depth of box in inches}$$

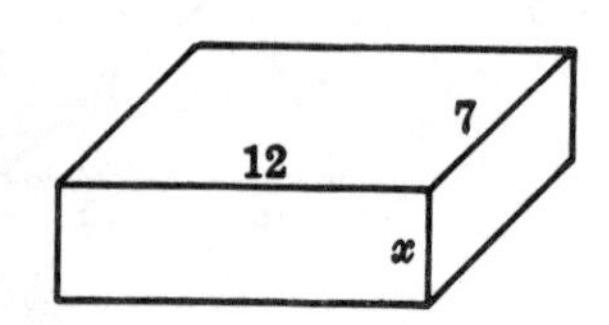

Equation

$$7(12)x = 336$$
$$84x = 336$$
$$x = 4 \quad \textit{Answer}$$

Check

$$(7)(12)(4) = 336$$
$$336 = 336$$

9. Let

$$x = \text{width of original rectangle in feet}$$
$$x + 8 = \text{length of original rectangle in feet}$$

	$x + 8$	
Original Rectangle	Area $x(x+8)$	x

Then

$$x + 4 = \text{width of new rectangle}$$
$$(x+8) - 5 = \text{length of new rectangle}$$

	$(x+8) - 5$	
New Rectangle	Area $(x+4)(x+3)$	$x + 4$

Equation

Area of first rectangle, $x(x+8)$, equals area of second rectangle, $(x+4)(x+3)$.

$$x(x+8) = (x+4)(x+3)$$
$$x^2 + 8x = x^2 + 7x + 12$$
$$8x - 7x = 12$$
$$x = 12$$
$$x + 8 = 20 \quad \textit{Answer}$$

Check

$$12(20) = (12+4)(12+3)$$
$$240 = 240$$

Chapter 10

Digits

Digit problems are problems involving numbers with two or more digits. In order to understand them, you must understand what each digit in a number means. For example, the number 329 means

3 one hundreds	or $3 \times 100 = 300$	
2 tens	or $2 \times 10 = 20$	
9 units (ones)	or $9 \times 1 = 9$	
	329	Sum

If you have a digit problem in algebra, you have to express the number as above. For example, if the tens digit is twice the units (ones) digit, we can express the number:

Let x = units digit

$2x$ = tens digit

Then if we multiply the tens digit by 10 and add it to the units digit, we get

$$\text{Number} = 10(2x) + x$$

Often a digit problem involves a *reversed* number; that is, the digits are interchanged. The numbers 39 and 93 have digits reversed. We can express these two numbers as follows: $(10 \times 3) + 9$ and $(10 \times 9) + 3$.

Now take the number above, $10(2x) + x$. To reverse it, we reverse the units digit x and the tens digit $2x$. We get $10(x) + 2x$.

EXAMPLE 1. The tens digit of a certain number is three less than the units digit. The sum of the digits is 11. What is the number?

Solution

Let

$$x = \text{units digit}$$
$$x - 3 = \text{tens digit}$$

Here there is no need to write the number itself except for the answer.

Equation

$$x + (x - 3) = 11$$
$$2x = 14$$
$$x = 7 \quad \text{units digit}$$
$$x - 3 = 4 \quad \text{tens digit}$$

Using the values above for units and tens, we find the number is $(4 \times 10) + 7$ or 47. *Answer*

EXAMPLE 2. The tens digit of a number is twice the units digit. If the digits are reversed, the new number is 27 less than the original. Find the original number.

Solution

Let

$$x = \text{units digit}$$
$$2x = \text{tens digit}$$

Original number is $10(2x) + x$.

Reversed number is $10(x) + 2x$.

New number is original number less 27.

Equation

$$10(x) + 2x = 10(2x) + x - 27$$
$$12x = 21x - 27$$
$$-9x = -27$$
$$x = 3 \quad \text{units digit}$$
$$2x = 6 \quad \text{tens digit}$$

The number is $(6 \times 10) + 3$ or 63. *Answer*

EXAMPLE 3. The sum of the digits in a two-digit number is 12. If the digits are reversed, the number is 18 greater than the original number. What is the number?

Solution

Let

$$x = \text{units digit}$$
$$12 - x = \text{tens digit}$$

Original number is $10(12 - x) + x$.

Reversed number is $10(x) + (12 - x)$.

Equation

Reversed number is original number plus 18 more.

$$10(x) + (12 - x) = 10(12 - x) + (x) + 18$$
$$9x + 12 = 120 - 10x + x + 18$$
$$9x + 12 = -9x + 138$$
$$18x = 126$$
$$x = 7 \quad \text{units digit}$$
$$12 - x = 5 \quad \text{tens digit}$$

The number is $(5 \times 10) + 7$ or 57. *Answer*

Check

$$10(7) + (12 - 7) = 10(12 - 7) + (7) + 18$$
$$70 + 5 = 50 + 7 + 18$$
$$75 = 75$$

SUPPLEMENTARY DIGIT PROBLEMS

1. The tens digit of a certain number is five more than the units digit. The sum of the digits is 9. Find the number.

2. The tens digit of a two-digit number is twice the units digit. If the digits are reversed, the new number is 36 less than the original number. Find the number.

3. The sum of the digits of a two-digit number is 13. The units digit is one more than twice the tens digit. Find the number.

4. The sum of the digits of a three-digit number is 6. The hundreds digit is twice the units digit, and the tens digit equals the sum of the other two. Find the number.

5. The units digit is twice the tens digit. If the number is doubled, it will be 12 more than the reversed number. Find the number.

6. Eight times the sum of the digits of a certain two-digit number exceeds the number by 19. The tens digit is three more than the units digit. Find the number.

7. The ratio of the units digit to the tens digit of a two-digit number is one-half. The tens digit is two more than the units digit. Find the number.

8. There is a two-digit number whose units digit is six less than the tens digit. Four times the tens digit plus five times the units digit equal 51. Find the digits.

9. The tens digit is two less than the units digit. If the digits are reversed, the sum of the reversed number and the original number is 154. Find the original number.

10. A three-digit number has a tens digit two greater than the units digit and the hundreds digit one greater than the tens digit. The sum of the tens and hundreds digits is three times the units digit. What is the number?

SOLUTIONS TO SUPPLEMENTARY DIGIT PROBLEMS

1. Let

$$x = \text{units digit}$$
$$x + 5 = \text{tens digit}$$

Equation

$$x + (x + 5) = 9$$
$$2x + 5 = 9$$
$$2x = 4$$
$$x = 2 \quad \text{units digit}$$
$$x + 5 = 7 \quad \text{tens digit}$$

Number is $(7 \times 10) + 2$ or 72. *Answer*

2. Let

$$x = \text{units digit}$$
$$2x = \text{tens digit}$$

Number is $10(2x) + x$.

Reversed number is $10(x) + 2x$.

Equation

$$10(x) + 2x = 10(2x) + x - 36$$
$$12x = 21x - 36$$
$$-9x = -36$$
$$x = 4 \quad \text{units digit}$$
$$2x = 8 \quad \text{tens digit}$$

The number is $(8 \times 10) + 4$ or 84. *Answer*

3. Let

$$x = \text{units digit}$$
$$13 - x = \text{tens digit}$$

Equation

Units digit is twice tens digit plus 1.

$$x = 2(13 - x) + 1$$
$$x = 26 - 2x + 1$$
$$3x = 27$$
$$x = 9 \quad \text{units digit}$$
$$13 - x = 4 \quad \text{tens digit}$$

The number is $(4 \times 10) + 9$ or 49. *Answer*

4. Let

$$x = \text{units digit}$$
$$2x = \text{hundreds digit}$$
$$x + 2x = \text{tens digit}$$

Equation

$$(x) + (2x) + (x + 2x) = 6$$
$$6x = 6$$
$$x = 1 \quad \text{units digit}$$
$$2x = 2 \quad \text{hundreds digit}$$
$$x + 2x = 3 \quad \text{tens digit}$$

The number is $(2 \times 100) + (3 \times 10) + 1$ or 231. *Answer*

5. Let

$$x = \text{tens digit}$$
$$2x = \text{units digit}$$

Number is $10(x) + (2x)$.

Reversed number is $10(2x) + (x)$.

Equation

Two times the number equals 12 more than the reversed number.

$$\begin{aligned} 2[10(x) + (2x)] &= 10(2x) + (x) + 12 \\ 2(12x) &= 21x + 12 \\ 24x &= 21x + 12 \\ 3x &= 12 \\ x &= 4 \quad \text{tens digit} \\ 2x &= 8 \quad \text{units digit} \end{aligned}$$

The number is $(4 \times 10) + 8$ or 48. *Answer*

6. Let x = units digit (smaller)

$x + 3$ = tens digit

Number is $10(x + 3) + x$.

Equation

Eight times the sum of the digits exceeds the number by 19.

$$\begin{aligned} 8[x + (x+3)] - [10(x+3) + x] &= 19 \\ 16x + 24 - 10x - 30 - x &= 19 \\ 5x - 6 &= 19 \\ 5x &= 25 \\ x &= 5 \quad \text{units digit} \\ x + 3 &= 8 \quad \text{tens digit} \end{aligned}$$

The number is $(8 \times 10) + 5 = 85$. *Answer*

7. Let x = units digit

$x + 2$ = tens digit

Equation

Ratio is a fractional relationship.

$$\frac{x}{x+2} = \frac{1}{2}$$

Multiply by LCD, $2(x+2)$.

$$2x = x + 2$$
$$x = 2 \quad \text{units digit}$$
$$x + 2 = 4 \quad \text{tens digit}$$

The number is $(4 \times 10) + 2$ or 42. *Answer*

8. Let

$$x = \text{units digit}$$
$$x + 6 = \text{tens digit}$$

Equation

Four times tens digit plus five times units digit equals 51.

$$4(x+6) + 5x = 51$$
$$4x + 24 + 5x = 51$$
$$9x + 24 = 51$$
$$9x = 27$$
$$x = 3 \quad \text{units digit}$$
$$x + 6 = 9 \quad \text{tens digit}$$

The number is $9 \times 10 + 3$ or 93. *Answer*

9. Let

$$x = \text{units digit}$$
$$x - 2 = \text{tens digit}$$

Number is $10(x-2) + x$.

Reversed number is $10(x) + (x-2)$.

Equation

Reversed number plus original number equals 154.

$$10(x) + (x-2) + 10(x-2) + x = 154$$
$$10x + x - 2 + 10x - 20 + x = 154$$
$$22x - 22 = 154$$
$$22x = 176$$
$$x = 8 \quad \text{units digit}$$
$$x - 2 = 6 \quad \text{tens digit}$$

The number is $(6 \times 10) + 8$ or 68. *Answer*

10. Let

$$x = \text{units digit}$$
$$x + 2 = \text{tens digit}$$
$$(x+2) + 1 = \text{hundreds digit}$$

Equation

The sum of the tens and hundreds digits is three times units digit.

$$(x+2) + (x+2) + 1 = 3x$$
$$2x + 5 = 3x$$
$$-x = -5$$
$$x = 5 \quad \text{units digit}$$
$$x + 2 = 7 \quad \text{tens digit}$$
$$(x+2) + 1 = 8 \quad \text{hundreds digit}$$

The number is $(8 \times 100) + (7 \times 10) + 5$ or 875. *Answer*

Chapter 11

Solutions Using Two Unknowns

This chapter will repeat examples of problems which were worked in earlier chapters. This time they will be worked using two unknowns (x and y) and two equations solved simultaneously. A few word problems are easier worked with one unknown, but many become easier when two unknowns are used. The same procedures that we learned in the problems with one unknown apply here also. You may find it easier to set up a problem with two unknowns because representing the second unknown is so easy. The first will be represented by x and the second by y. Remember, *values must be paired at the end of the problem.* The value of x alone is not enough.

The problems which follow are grouped to correspond to the appropriate earlier chapters. Examples of most types of problems will be worked again with two unknowns. Occasionally, a certain type of problem is more difficult with two unknowns.

NUMBER PROBLEMS

1. The sum of two numbers is 41. The larger number is one less than twice the smaller. Find the numbers.

Solution

The last sentence tells us we have two unknowns. Now we shall use x and y to represent these unknowns. The two statements in the problem give us information for two equations.

Let

$$x = \text{smaller number}$$
$$y = \text{larger number}$$

Equations

The sum of two numbers is 41.

$$(1) \quad x + y = 41$$

The larger is one less than twice the smaller.

$$(2) \quad y = 2x - 1$$

Solve the two equations together.

$$(1)\ x + y = 41$$

$$(2)\ y = 2x - 1$$

Substitute value of y in (2) for y in (1) because (2) is already solved for y, making substitution easy.

$$\begin{aligned} (1) \quad x + (2x - 1) &= 41 \\ 3x - 1 &= 41 \\ 3x &= 42 \\ x &= 14 \end{aligned}$$

Substituting x in (2),

$$\begin{aligned} (2) \quad y &= 2x - 1 \\ y &= 2(14) - 1 \\ y &= 27 \end{aligned}$$

$$\left.\begin{aligned} x &= 14 \\ y &= 27 \end{aligned}\right\} \quad \textit{Answers}$$

When you use the substitution method, remember to substitute the value of the first unknown in the equation which was used in the substitution step to get the second unknown. In this problem you substitute the value of x back into $y = 2x - 1$.

2. There are two numbers whose sum is 53. Three times the smaller number is equal to 19 more than the larger number. What are the numbers?

Solution

Let x = smaller number

y = larger number

Equations

Two numbers whose sum is 53.

$$(1) \quad x + y = 53$$

Three times the smaller equals 19 more than larger.

$$(2) \quad 3x = y + 19$$

$$\begin{aligned} (1) \quad x + y &= 53 \\ (2) \quad 3x - y &= 19 \end{aligned}$$

Add the two equations to eliminate y.

$$4x = 72$$

$$x = 18$$

Substituting x in (1),

$$(1)\quad x + y = 53$$

$$18 + y = 53$$

$$y = 35$$

$$\left.\begin{aligned} x &= 18 \\ y &= 35 \end{aligned}\right\} \quad \textit{Answers}$$

TIME, RATE, AND DISTANCE PROBLEMS

1. Kevin flies (in his private plane) from Los Angeles to San Francisco, a distance of 450 miles, in 3 hours. He flies back in $2\frac{7}{19}$ hours. If the wind is blowing from the north at a steady rate during both flights, find the airspeed of the plane and the velocity of the wind.

Read the problem carefully. There are two unknowns. There are two statements of relationship, one for each equation. Determine what you are trying to find. You always represent the two unknowns by two different letters in this type of solution. The basic procedures are the same as with one unknown. Draw the sketch:

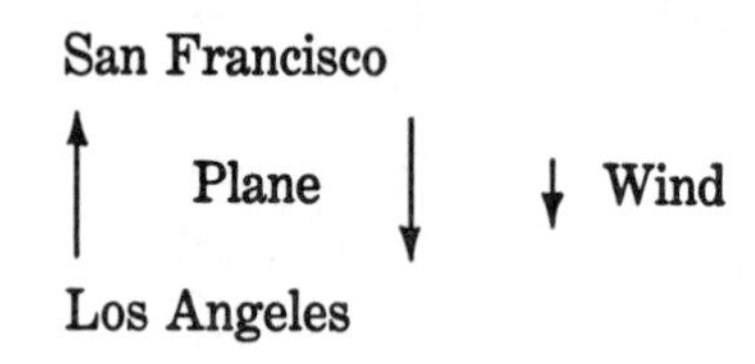

Solution

Let x = airspeed of plane in mph (in still air!)

y = velocity of wind

	Time	Rate	Distance
LA to SF	3	$x - y$	$3(x - y)$
SF to LA	$2\frac{7}{19}$	$x + y$	$2\frac{7}{19}(x + y)$

Here there are *three* possible equations. You can use any two.

$$(1)\quad 3(x - y) = 450$$

$$(2)\quad 2\tfrac{7}{19}\,(x + y) = 450$$

$$(3)\quad 3(x - y) = 2\tfrac{7}{19}\,(x + y)$$

It is more convenient to use equations (1) and (2) because the unknowns are on the same side of the equation (proper format for equations in 2 unknowns).

$$(1)\quad 3(x - y) = 450$$

$$(2)\quad 2\tfrac{7}{19}\,(x + y) = 450$$

Clear parentheses:

$$(1)\quad 3x - 3y = 450$$

$$(2)\quad \frac{45x + 45y}{19} = 450$$

Clear fraction:

$$(1)\quad 3x - 3y = 450$$

$$(2)\quad 45x + 45y = 8550$$

Divide equation (1) by 3 and equation (2) by 45:

$$(1)\quad x - y = 150$$

$$(2)\quad x + y = 190$$

Add equations and solve for x:

$$2x = 340$$

$x = 170$ mph, airspeed *Answer*

Substitute x in (2):

$$(2)\quad x + y = 190$$

$$170 + y = 190$$

$y = 20$ mph, wind velocity *Answer*

MIXTURE PROBLEMS

1. A farmer has 100 gallons of 70% pure disinfectant. He wishes to mix it with disinfectant which is 90% pure to obtain 75% pure disinfectant. How much of the 90% must he use?

Solution

There are two unknowns. Either one may be labeled x and the other y. Note that we only have to find one in this problem.

Let

$$x = \text{number of gallons of 90\% disinfectant}$$
$$y = \text{number of gallons of 75\% disinfectant}$$

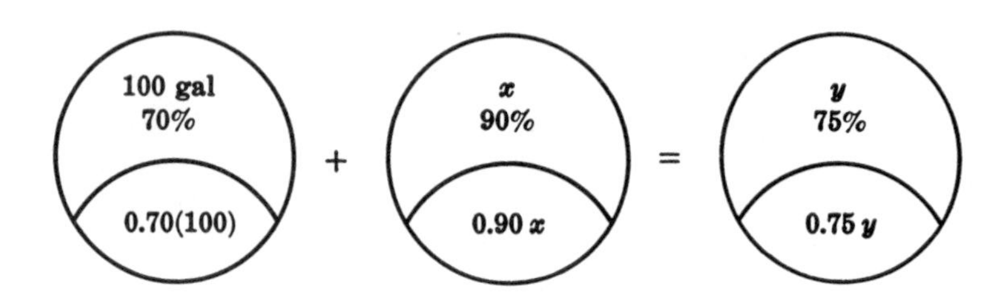

Equations

(1) $100 + x = y$ (total gallons)

(2) $0.70(100) + 0.90\,x = 0.75\,y$ (total disinfectant)

(2) $70 + 0.90\,x = 0.75\,y$

$7000 + 90x = 75y$

Divide by 5.

(2) $1400 + 18x = 15y$

Substitute (1) in (2).

$$1400 + 18x = 15(100 + x)$$
$$1400 + 18x = 1500 + 15x$$
$$3x = 100$$
$$x = 33\tfrac{1}{3} \text{ gallons} \quad \textit{Answer}$$

2. A farmer mixes milk containing 3% butterfat with cream containing 30% butterfat to obtain 900 gallons of milk which is 8% butterfat. How much of each must he use?

Solution

Let x = number of gallons of milk

y = number of gallons of cream

Equations

$$(1) \quad x + y = 900$$

$$(2) \quad 0.03\,x + 0.30\,y = 0.08(900)$$

$$0.03\,x + 0.30\,y = 72$$

Multiply by 100 to clear decimals.

$$(2) \quad 3x + 30y = 7200$$

Divide equation (2) by 3.

$$(2) \quad x + 10y = 2400$$

$$(1) \quad x + y = 900$$

Subtract equation (1) from (2) to eliminate x.

$$9y = 1500$$

$y = 166\frac{2}{3}$ gallons of 30% butterfat

Substitute y in (1)

$x = 733\frac{1}{3}$ gallons of 3% butterfat

} *Answers*

COIN PROBLEMS

1. Ezekiel has some coins in his pocket consisting of dimes, nickels, and pennies. He has two more nickels than dimes and three times as many pennies as nickels. How many of each kind of coin does he have if the total value is 52¢?

Solution

Let

x = number of dimes

y = number of nickels

z = number of pennies

Equations

$$(1) \quad y = x + 2$$

$$(2) \quad z = 3y$$

$$(3) \quad 10x + 5y + z = 52$$

Remember to convert x, y, and z to cents in equation (3). Then substitute (1) and (2) in (3) so all is in terms of y.

$$(3) \quad 10(y-2) + 5y + (3y) = 52$$

$$18y = 72$$

$$\left.\begin{aligned} y &= 4 \text{ nickels} \\ z = 3y \quad \text{so} \quad z &= 12 \text{ pennies} \\ x = y - 2 \quad \text{so} \quad x &= 2 \text{ dimes} \end{aligned}\right\} \quad \textit{Answers}$$

2. Mrs. Delahoyd received a total of 180 coins each day for her cash drawer at the restaurant where she was a cashier. She had the same number of nickels as quarters and the same number of pennies as dimes. She had ten more dimes than nickels. How many of each did she have?

Solution

Let

P = number of pennies

N = number of nickels

D = number of dimes

Q = number of quarters

Equations

$$(1) \quad P + N + D + Q = 180$$

$$(2) \quad N = Q$$

$$(3) \quad P = D$$

$$(4) \quad D = N + 10$$

We need to find two equations which will each have the same two unknowns. First, substitute (2) and (3) in (1) to make an equation with values in Q and D.

$$(1) \quad (D) + (Q) + D + Q = 180$$

$$2D + 2Q = 180$$

Divide by 2.

$$(1) \quad D + Q = 90$$

Second, substitute (2) into (4) to get a second equation with values in Q and D.

$$(4) \quad D = (Q) + 10$$

Solve (1) and (4) together.

$$(1) \quad D + Q = 90$$

$$(4) \quad D - Q = 10$$

Add equations (1) and (4) to eliminate Q.

$$2D = 100$$
$$D = 50$$

Substitute value for D in equation (1) to find Q.

$$(1) \quad D + Q = 90$$
$$50 + Q = 90$$
$$Q = 40$$

Substitute value for Q in equation (2) to find N.

$$(2) \quad N = Q$$
$$N = 40$$

Substitute value for D in equation (3) to find P.

$$(3) \quad P = D$$
$$P = 50$$

$$\left.\begin{aligned} D &= 50 \\ Q &= 40 \\ N &= 40 \\ P &= 50 \end{aligned}\right\} \quad \textit{Answers}$$

AGE PROBLEMS

1. A man is four times as old as his son. In 3 years the father will be three times as old as the son. How old is each now?

Solution

Let

$$x = \text{son's age now}$$
$$y = \text{father's age now}$$
$$x + 3 = \text{son in 3 years}$$
$$y + 3 = \text{father in 3 years}$$

Equations

Father's age is 4 times son's age *now*.

$$(1) \quad y = 4x$$

Father's age is 3 times son's age *then*.

$$(2) \quad y + 3 = 3(x + 3)$$

Substitute (1) in (2).

$$(2)\quad (4x) + 3 = 3(x+3)$$
$$4x + 3 = 3x + 9$$
$$x = 6$$

Substituting x in (1),

$$\left.\begin{aligned} y &= 24 \\ x &= 6 \end{aligned}\right\} \quad \textit{Answers}$$

2. Bettina's age is three times Melvina's. If 20 is added to Melvina's age and 20 is subtracted from Bettina's, their ages will then be equal. How old is each now?

Solution

Let x = Bettina's age now

y = Melvina's age now

Equations

Bettina's age is three times Melvina's.

$$(1)\quad x = 3y$$

If 20 is added to Melvina's age and 20 is subtracted from Bettina's, their ages are equal.

$$(2)\quad y + 20 = x - 20$$

Substitute (1) in (2).

$$(2)\quad y + 20 = (3y) - 20$$
$$-2y = -40$$
$$y = 20$$

Substitute y in (1).

$$(1)\quad x = 3y$$
$$x = 3(20)$$
$$\left.\begin{aligned} x &= 60 \\ y &= 20 \end{aligned}\right\} \quad \textit{Answers}$$

LEVER PROBLEMS

Lever problems are easier in one unknown. In general, you will find some problems easier to work in one unknown and some easier in two unknowns. There is no rule to follow.

FINANCE PROBLEMS

1. Mr. McGregor invested \$10,000, part at 5% and part at 9%. If the total yearly interest was \$660, how much did he invest at each rate?

Solution

Let x = amount in dollars invested at 5%

y = amount in dollars invested at 9%

$0.05\,x$ = interest on 5% investment

$0.09\,y$ = interest on 9% investment

Equations

(1) $0.05\,x + 0.09\,y = 660$ (Total *interest* equals \$660)

(2) $x + y = 10{,}000$ (Total *investment* equals \$10,000)

Multiply equation (1) by 100 and multiply equation (2) by 5.

(1) $5x + 9y = 66{,}000$

(2) $5x + 5y = 50{,}000$

Subtract to eliminate x.

$$4y = 16{,}000$$

$$y = 4000$$

Substitute y in equation (2).

(2) $x + 4000 = 10{,}000$

$$x = 6000$$

Always pair x and y values to show solution set.

$$\left.\begin{aligned} x &= 6000 \\ y &= 4000 \end{aligned}\right\} \quad \textit{Answers}$$

2. Tickets for the school play sold at \$2 each for adults and 75¢ each for students. If there were four times as many adult tickets sold as student tickets, and the total receipts were \$1750, how many adult and how many student tickets were sold?

Solution

Let

$$x = \text{number of adult tickets sold at \$2 each}$$

$$y = \text{number of student tickets sold at 75¢ each}$$

$$2x = \text{total number of dollars received from adult tickets}$$

$$0.75y = \text{total number of dollars received from student tickets}$$

Equations

Four times as many adult tickets as student tickets sold.

$$(1) \quad x = 4y$$

Total receipts were $1750.

$$(2) \quad 2x + 0.75y = 1750$$

Clear decimals in (2).

$$(2) \quad 200x + 75y = 175{,}000$$

Divide (2) by constant 25 to simplify equation.

$$(2) \quad 8x + 3y = 7000$$

Substitute value of x in (1) into equation (2) so that equation (2) will all be in terms of y.

$$(2) \quad 8(4y) + 3y = 7000$$

$$35y = 7000$$

$$y = 200$$

Substitute y in (1).

$$(1) \quad x = 4y$$

$$x = 4(200)$$

$$x = 800$$

$$\left.\begin{aligned} x &= 800 \\ y &= 200 \end{aligned}\right\} \textit{Answers}$$

WORK PROBLEMS

Work problems are more easily solved with one unknown.

PROBLEMS INVOLVING PLANE GEOMETRIC FIGURES

1. The length of a rectangle is 5 feet more than twice the width. The perimeter is 28 feet. Find the length and width of the rectangle.

Solution

Let x = width in feet

y = length in feet

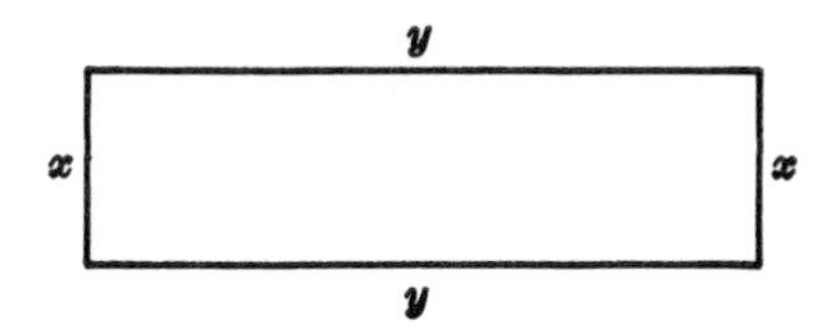

Equations

Length is 5 more than twice the width.

$$(1) \quad y = 2x + 5$$

Perimeter is 28 feet.

$$(2) \quad 2x + 2y = 28$$

Divide equation (2) by constant 2 to simplify.

$$(2) \quad x + y = 14$$

Substitute value of y in (1) into (2) to express equation (2) in terms of x.

$$(2) \quad x + (2x + 5) = 14$$
$$3x + 5 = 14$$
$$3x = 9$$
$$x = 3$$

Substitute x in equation (1).

$$(1) \quad y = 2(3) + 5$$
$$y = 6 + 5$$
$$y = 11$$

$$\left.\begin{aligned} x &= 3 \\ y &= 11 \end{aligned}\right\} \quad \textit{Answers}$$

2. The second angle of a triangle is 20 degrees greater than the first angle. The third is twice the second. Find the three angles.

Solution

Let x = number of degrees in first angle

y = number of degrees in second angle

z = number of degrees in third angle

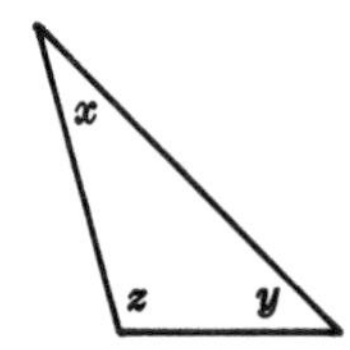

Equations

The second angle is 20 degrees greater than the first angle.

$$(1) \quad y = x + 20$$

The third is twice the second.

$$(2) \quad z = 2y$$

The sum of the angles of a triangle is 180 degrees.

$$(3) \quad x + y + z = 180$$

With three unknowns there must be three equations.

Remember, we need to get two equations in two unknowns to solve them together to eliminate one unknown. Sometimes it is possible to substitute two equations into the third so that all but one unknown is eliminated. We can solve equation (1) for x. Equation (2) has z in terms of y. So we can now substitute (1) and (2) into equation (3) to get (3) in terms of y.

$$(1) \quad x = y - 20$$

$$(2) \quad z = 2y$$

$$(3) \quad x + y + z = 180$$

$$(3) \quad (y - 20) + y + (2y) = 180$$

$$4y - 20 = 180$$
$$4y = 200$$
$$y = 50$$
$$z = 2y$$
$$z = 100$$
$$x = y - 20$$
$$x = 30$$

$$\left.\begin{aligned} x &= 30 \\ y &= 50 \\ z &= 100 \end{aligned}\right\} \quad \textit{Answers}$$

DIGIT PROBLEMS

1. The tens digit of a two-digit number is five more than the units digit. The sum of the digits is 9. Find the number.

Solution

Let x = units digit

y = tens digit

You could let x = tens and y = units. In using two letters for the two unknowns, it does not make any difference which unknown x stands for and which unknown y stands for. You will get the correct answer either way.

The tens digit is 5 more than the units digit.

$$(1) \quad y = x + 5$$

The sum of the digits is 9.

$$(2) \quad x + y = 9$$

Because equation (1) is already solved for y in terms of x, it suggests substitution. You substitute $x + 5$ for y in equation (2).

$$(2) \quad x + (x + 5) = 9$$
$$2x + 5 = 9$$
$$2x = 4$$
$$x = 2$$

To find the value of y, go back to equation (1) which expresses y in terms of x.

$$\text{(1)}\quad y = x + 5$$
$$y = (2) + 5$$
$$y = 7$$
$$\left.\begin{aligned} x &= 2 \\ y &= 7 \end{aligned}\right\}$$

The number is 72. *Answer*

2. The sum of the digits of a three-digit number is 6. The hundreds digit is twice the units digit. The tens digit equals the sum of the other two digits. What is the number?

Solution

Let

$$u = \text{units digit}$$
$$t = \text{tens digit}$$
$$h = \text{hundreds digit}$$

The sum of the digits is 6.

$$\text{(1)}\quad u + t + h = 6$$

The hundreds digit is twice the units digit.

$$\text{(2)}\quad h = 2u$$

The tens digit equals the sum of the other two.

$$\text{(3)}\quad t = u + h$$

Look at the above equations to see what relationships you have. (2) has h in terms of u, so see if you can find another letter in terms of u. If you substitute (2) into (3) you will have a t in terms of u.

$$\text{(3)}\quad t = u + (2u)$$
$$\text{(3)}\quad t = 3u$$

So you now have

$$\text{(1)}\quad u + t + h = 6$$
$$\text{(2)}\quad h = 2u$$
$$\text{(3)}\quad t = 3u$$

By substituting (2) and (3) in (1) all will be in terms of u.

$$\text{(1)}\quad u + (3u) + (2u) = 6$$
$$6u = 6$$
$$u = 1$$

$$h = 2u \qquad h = 2$$
$$t = 3u \qquad t = 3$$

The number is 231. *Answer*

Chapter 12

Quadratics

Certain problems of various types involve quadratic equations. However, as in solving equations with two letters representing the unknowns (Chapter 11), the basic procedures remain the same.

EXAMPLE 1. A rectangular pool is surrounded by a walk 4 feet wide. The pool is 6 feet longer than it is wide. If the total area is 272 square feet more than the area of the pool, what are the dimensions of the pool?

Solution

Let $\qquad x$ = width of pool in feet

$\qquad x + 6$ = length of pool in feet

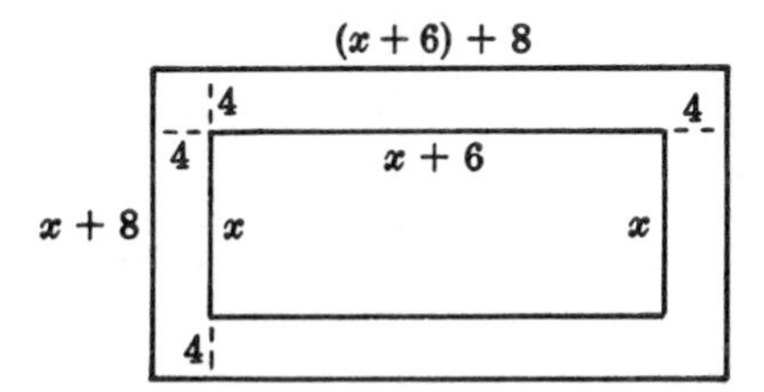

The dimensions of the outside rectangle will each be 8 feet more than the dimensions of the inside rectangle (two widths of the walk).

Area of pool is $x(x + 6)$.

Area of large rectangle (total area) is $(x + 8)(x + 14)$.

Equation

Total area is 272 square feet more than pool area.

$$(x+8)(x+14) = x(x+6) + 272$$
$$x^2 + 22x + 112 = x^2 + 6x + 272$$
$$16x = 160$$
$$\left.\begin{aligned} x &= 10 \\ x+6 &= 16 \end{aligned}\right\} \quad \textit{Answers}$$

EXAMPLE 2. One man can lay a cement slab in 2 hours less than it takes his competitor to do the same job. If they worked together, they could do it in $2\frac{2}{5}$ hours. How long does it take each to lay the slab alone?

Solution

This is a typical work problem, so you diagram it as in Chapter 8.

Let x = time in hours for first man

$x + 2$ = time in hours for second man

	Total time in hours	Fractional part of job done in 1 hour
First man	x	$\frac{1}{x}$
Second man	$x+2$	$\frac{1}{x+2}$
Together	$2\frac{2}{5}$	$\frac{1}{2\frac{2}{5}}$

Equation

Total fractional part each can do in 1 hour equals fractional part they can do together.

$$\frac{1}{x} + \frac{1}{x+2} = \frac{1}{2\frac{2}{5}}$$

$$\frac{1}{2\frac{2}{5}} = \frac{1}{\frac{12}{5}} = 1 \div \frac{12}{5} = \frac{5}{12}$$

$$\frac{1}{x} + \frac{1}{x+2} = \frac{5}{12}$$

Multiply by LCD, $12x(x+2)$, to clear fractions.

$$
\begin{aligned}
12(x+2) + 12x &= 5x(x+2) \\
12x + 24 + 12x &= 5x^2 + 10x \\
-5x^2 - 10x &= -24x - 24 \\
5x^2 + 10x &= 24x + 24 \\
5x^2 - 14x - 24 &= 0 \\
(5x+6)(x+4) &= 0 \\
x - 4 &= 0 \\
\left.\begin{aligned} x &= 4 \\ x + 2 &= 6 \end{aligned}\right\} &\quad \textit{Answers}
\end{aligned}
$$

The second factor gives a negative root, which is unacceptable in a word problem.

EXAMPLE 3. Jason traveled 120 miles at an average rate of speed. If he could have increased his speed 15 mph, he could have covered the same distance in $\frac{2}{5}$ of an hour less time. How fast did he travel?

Solution

Let $\quad x$ = rate in mph before increase

$x + 15$ = rate in mph after increase

	Time	Rate	Distance
Before	$\frac{120}{x}$	x	120
After	$\frac{120}{x + 15}$	$x + 15$	120

Equation

Time *before* increase is $\frac{2}{5}$ of an hour more than time after.

$$\frac{120}{x} = \frac{120}{x + 15} + \frac{2}{5}$$

Multiply by LCD to clear fractions.

$$
\begin{aligned}
5(120)(x + 15) &= 5(120)x + 2x(x + 15) \\
600x + 9000 &= 600x + 2x^2 + 30x \\
2x^2 + 30x - 9000 &= 0 \\
x^2 + 15x - 4500 &= 0 \\
(x + 75)(x - 60) &= 0
\end{aligned}
$$

$$x + 75 = 0$$

$$x = -75 \text{ (drop as extraneous answer)}$$

$$x - 60 = 0$$

$$\left.\begin{aligned} x &= 60 \text{ mph for trip before increase} \\ x + 15 &= 75 \text{ mph if speed increased} \end{aligned}\right\} \textit{Answers}$$

NOTE: Remember that in quadratics you will sometimes have an extra answer which is negative and has to be discarded. When you substitute to find second answer, it *sometimes* turns out to be a positive form of this number. Don't count on it!

EXAMPLE 4. Two consecutive even numbers have a product of 624. What are the numbers?

Solution

Let x = first consecutive even number

$x + 2$ = second consecutive even number

Equation

$$x(x+2) = 624$$

$$x^2 + 2x = 624$$

$$x^2 + 2x - 624 = 0$$

$$(x+26)(x-24) = 0$$

$$x - 24 = 0$$

$$\left.\begin{aligned} x &= 24 \text{ first consecutive even number} \\ x + 2 &= 26 \text{ second consecutive even number} \end{aligned}\right\} \textit{Answers}$$

EXAMPLE 5. A 60 by 80 foot rectangular walk in a park surrounds a flower bed. If the walk is of uniform width and its area is equal to the area of the flower bed, how wide is the walk?

Solution

Let x = width of walk in feet

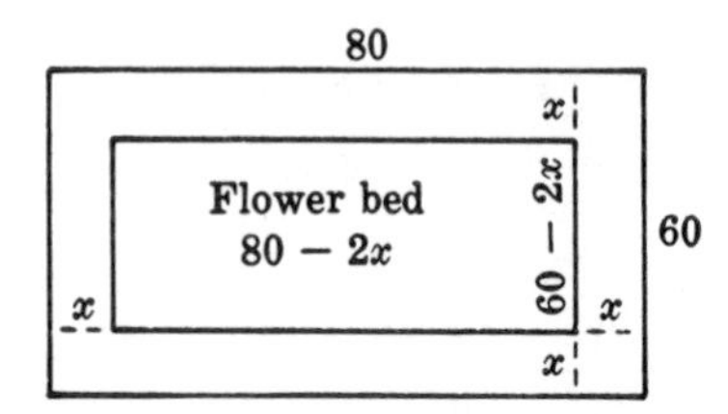

Area of large rectangle is $80 \times 60 = 4800$.

Area of small rectangle is $\frac{1}{2}$ area of large rectangle $= 2400$.

Area of small rectangle is $(80-2x)(60-2x)$.

Equation

$$(80-2x)(60-2x) = 2400$$
$$4800 - 280x + 4x^2 = 2400$$
$$4x^2 - 280x + 2400 = 0$$
$$x^2 - 70x + 600 = 0$$
$$(x-10)(x-60) = 0$$
$$x - 10 = 0$$
$$x = 10, \text{ width of walk in feet} \qquad \textit{Answer}$$
$$x - 60 = 0$$
$$x = 60, \text{ width of walk in feet}$$

(not acceptable; too large for width of sidewalk)

EXAMPLE 6. A girl is 12 years older than her sister. The product of their ages is 540. How old is each?

Solution

Let x = sister's age

$x + 12$ = girl's age

Equation

$$x(x+12) = 540$$
$$x^2 + 12x - 540 = 0$$
$$(x+30)(x-18) = 0$$
$$x - 18 = 0$$
$$\left.\begin{aligned} x &= 18 \text{ yrs.} \\ x + 12 &= 30 \text{ yrs.}\end{aligned}\right\} \begin{aligned}&\text{sister's age}\\ &\text{girl's age}\end{aligned} \qquad \textit{Answers}$$
$$x + 30 = 0$$
$$x = -30 \qquad \text{unacceptable}$$

You may have noticed that frequently the absolute value of the negative root is equal to the absolute value of the second unknown. Don't depend on it, however. It is not always true!

EXAMPLE 7. A number is one more than twice another. Their squares differ by 176. What are the numbers?

Solution

Let x = smaller number

$2x + 1$ = larger number

Equation

$$(2x+1)^2 = x^2 + 176$$
$$4x^2 + 4x + 1 = x^2 + 176$$
$$3x^2 + 4x - 175 = 0$$
$$(3x+25)(x-7) = 0$$
$$x - 7 = 0$$
$$x = 7 \quad \text{smaller number}$$
$$2x + 1 = 15 \quad \text{larger number}$$

Answers

EXAMPLE 8. The side of a square equals the width of a rectangle. The length of the rectangle is 6 feet longer than its width. The sum of their areas is 176 square feet. Find the side of the square.

Solution

Let x = side of square

$x + 6$ equals length of rectangle.

x^2 equals area of square.

$x(x + 6)$ equals area of rectangle.

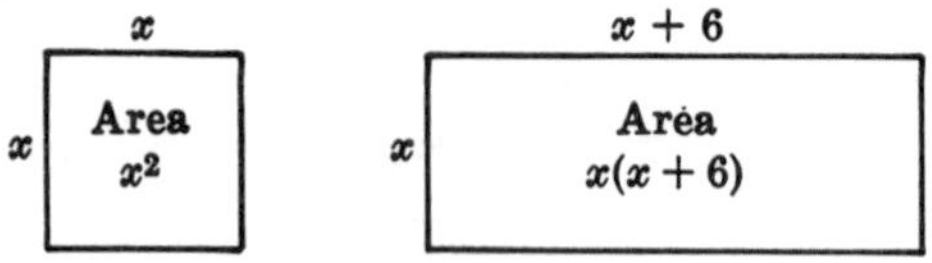

x | $x + 6$
x Area x^2 | x Area $x(x+6)$

Equation

$$x^2 + x(x+6) = 176$$
$$x^2 + x^2 + 6x = 176$$
$$2x^2 + 6x - 176 = 0$$
$$x^2 + 3x - 88 = 0$$
$$(x+11)(x-8) = 0$$
$$x - 8 = 0$$
$$x = 8, \text{ side of square in feet}$$

Answer

Chapter 13

Miscellaneous Problem Drill

1. One number is 11 more than three times another. Their sum is 111. What are the numbers?

2. The denominator of a fraction is 24 more than the numerator. The value of the fraction is $\frac{1}{3}$. Find the numerator and denominator.

3. There are three consecutive integers. The sum of the smallest and largest is 36. Find the integers.

4. Take a number. Double the number. Subtract 6 from the result and divide the answer by 2. The quotient will be 20. What is the number?

5. The sum of three consecutive odd numbers is 249. Find the numbers.

6. There are three consecutive even numbers such that twice the first is 20 more than the second. Find the numbers.

7. A carpenter needs to cut a 14-foot board into three pieces so that the second piece is twice as long as the first and the third is twice the second. How long is the shortest piece?

8. Twice a certain number plus three times the same number is 135. Find the number.

9. A two-digit number has a tens digit one greater than the units digit. The sum of the number and the number formed by reversing the digits is 77. Find the number.

10. The tens digit of a two-digit number is five more than the units digit. If 3 is subtracted from the number and 2 is added to the reversed number, the former will be twice the latter. What is the number?

11. Mr. Geld has $50,000 to invest. Part of it is put in the bank at 6%, and part he puts in a savings and loan at 9%. If his yearly interest (simple) is $3660, how much did he invest at each rate?

12. A store advertises a 20%-off sale. If an article is marked for the sale at $24.48, what is the regular price?

13. The Mountaineering Shop was owned jointly by Dave, Steven, and Pierre. Steve put up $2000 more than Dave, and Pierre owned a half interest. If the total cost of the shop was $52,000, how much did each man invest?

14. Three houses were for rent by the Jippem Realty Company. They charged $300 more per month for the second house than for the first. The third house rented for the same as the second but was vacant for two months for repairs. How much per month did each house rent for if the rent receipts for the year were $37,200?

15. Wolfgang and Heinrich worked as electricians for $20 and $22 per hour respectively. One month Wolfgang worked 10 hours more than Heinrich. If their total income for the month was $5660, how many hours did each work during the month?

16. Mike and Ike each inherited a sum of money from an uncle. Mike received $800 more than Ike. Ike invested his at 8% and Mike invested his at 7%. If Mike received $16 a year more than Ike in interest, how much did each inherit?

17. Mary needs a 50% solution of alcohol. How many liters of pure alcohol must she add to 10 liters of 40% alcohol to get the proper solution?

18. A 20% nitric acid solution and a 45% nitric acid solution are to be mixed to make 6 quarts 30% acid. How much of each must be used?

19. Jones has a 90% solution of boric acid in his pharmacy which he reduces to the required strength by adding distilled water. How much solution and how much water must he use to get 2 quarts of 10% solution?

20. A service station checks Mr. Gittleboro's radiator and finds it contains only 30% antifreeze. If the radiator holds 10 quarts and is full, how much must be drained off and replaced with pure antifreeze in order to bring it up to a required 50% antifreeze?

21. The Dingles have some friends drop in. They wish to serve sherry but do not have enough to serve all the same kind. So they mix some which is 20% alcohol with some which is 14% alcohol and have 5 quarts which is 16% alcohol. How much of each kind of sherry did they have to mix?

22. A bus leaves Riverside for Springfield, and averages only 48 mph. Ten minutes later Smith leaves Riverside for Springfield traveling 64 mph in his car. How long before he overtakes the bus?

23. Two trains leave Chicago, one headed due east at 60 mph and one headed due west at 50 mph. How long before they will be 990 miles apart?

24. A scenic road around Lake Rotorua is 11 miles in length. Tom leaves the hotel on his bike averaging ten mph and heads west around the lake. Bill leaves at the same time and heads in the opposite direction at 45 mph. How far from the hotel will Tom have traveled when they meet?

25. A small boat sends a distress signal giving its location as 10 miles from shore in choppy seas and says it is making only 5 knots. A Coast Guard boat is dispatched from shore to give aid; it averages 20 knots. How long before it meets the disabled boat if they are traveling toward each other? How far did each travel?

26. The Allisons are on a cross-country trip traveling with the Jensons. One day they get separated and the Jensons are 20 miles ahead of the Allisons on the same road. If the Jensons average 50 mph and the Allisons travel 60 mph, how long before the Allisons catch up with the Jensons?

27. Three-fifths of the men in a chemistry class have beards and two-thirds of the women have long hair. If there are 120 in the class and 46 are not in the above groups, how many men and how many women are there in the class?

28. The Girl Scouts have a yearly cookie sale. One year they have two varieties which don't sell so well, so they decide to mix them. The coconut macaroons sell for $3 a pound and the maple dates sell for $4 a pound. How many pounds of each do they use so that they will make the same amount of money but have 100 pounds of mixture and sell it at $3.20 per pound?

29. Mr. Higglebotham traveled 60 miles across the rainy English countryside at a constant rate of speed. If it had been sunny, he could have averaged 20 mph more and arrived at Broadmoor in 30 minutes less time. How fast was he driving? (This problem takes a knowledge of quadratics.)

30. A plane flies from Los Angeles west to Paradise Isle and returns. During both flights there is a steady upper air wind from the west at 80 miles per hour. If the trip west to P.I. took 17 hours and the return trip to L.A. took 13 hours, what was the plane's average air speed?

31. Timmy and Janie sit on opposite ends of a 20-foot teeter-totter. Timmy weighs 60 pounds and Janie weighs 40 pounds. Where would Sally, who weighs 50 pounds, have to sit to balance the teeter-totter? (Fulcrum at center.)

32. Jerry needs to pry a 50-pound rock out of his garden. If he uses a 6-foot lever and rests it on a board 1 foot from the rock (for the fulcrum), how much force must he exert to raise the rock? (Assume balance.)

33. Where must the fulcrum be located if a 250-pound weight and a 300-pound weight balance when placed on each end of an 11-foot bar?

34. An 88-pound boy sits on one end of a 15-foot board, 4 feet from the point of balance. His friend comes along and gets on the other side at a position which enables them to balance. How far from the fulcrum will the friend be if he weighs 64 pounds?

35. Tim and Tom sit on opposite ends of a 20-foot seesaw. If Tim weighs 120 pounds and is 8 feet from the fulcrum, how much does Tom weigh if they balance?

36. Mr. Swanson needs to move a 350-pound refrigerator. He has no dolly, so he gets his son to help balance the refrigerator while he slides a 6-foot board under it and uses a block as fulcrum. (This enables him to get a rug under the refrigerator on which it will slide.) If the fulcrum is 18 inches from the end of the board, how much force is needed to raise the refrigerator?

37. Tickets for the local baseball game were $15 for general admission and $8 for kids. There were 20 times as many general admission tickets sold as there were kids tickets. Total receipts were $616,000. How many of each type ticket were sold?

38. Phineas has $1.15 worth of change in his pocket. He has three more dimes than quarters and two more dimes than nickels. How many of each type of coin has he?

39. Bob has a coin collection made up of pennies and nickels. If he has three times as many pennies as nickels and the total face value of the coins is $16, how many coins of each kind are in the collection?

40. The boys have a small game going in the back room. Mr. X decides to pull out and finds he has $194. If he has four times as many $1 bills as $5 bills, one-fifth as many $2 as $5 bills, and the same number of $50 bills as $2 bills, how many bills of each kind does he have?

41. Jake Zablowski has a collection of coins worth $96.07. He has 5 times as many 50 cent pieces as he has silver dollars. The number of dimes is twice the number of 50 cent pieces. There are seven more than 3 times as many pennies as dimes. How many of each kind of coin does he have?

42. Zac is ten years older than Zelda. Twelve years ago he was twice as old. How old is each now?

43. Cassandra is twice as old as Zitka. If 12 is added to Cassandra's age and 6 is subtracted from Zitka's, Cassandra will be six times as old. How old is each now?

44. Maggie and Rod left at 9 AM on a hike to Tahquitz peak, traveling at an average rate of 2 mph, with 3 five-minute rest stops. After a 45-minute lunch at the peak, they returned home averaging 3 mph and arrived at 3 PM. How long did it take them to reach the peak and how *far* did they hike to reach it?

45. Jay's father is twice as old as Jay. In 20 years Jay will be two-thirds as old as his father. How old is each now?

46. Abigail will not reveal her age but says she is 3 years younger than her sister Kate. Ninety years ago Kate was twice as old. How old is each now?

47. Chuck is 22 years older than Jack. When Jack is as old as Chuck is now, he will be three times his present age. How old is each now?

48. The length of a garden is 3 feet more than four times the width. The fence enclosing it is 36 feet long. What are the dimensions?

49. A farmer wishes to fence a circular ring for his ponies. If he has 264 feet of fencing, what would he make the radius of the circle? (Use $\pi = \frac{22}{7}$.)

50. The first angle of a triangle is twice the second. The third is 5 degrees greater than the first. Find the angles.

51. Mr. Smith and Mr. Jones each built a similar stone fence to enclose his back yard. They only enclosed three sides, each using his house as the fourth side and had a community fence between the two yards. If the total cost was $5 per linear foot, the yards were 20 feet wider than they were deep, and the total cost was $950, what was the length and width of each yard?

52. The length of a certain rectangle is 4 feet greater than the width. If the length is decreased by 3 and the width is increased by 5, the area will increase by 21 square feet. What were the dimensions of the rectangle?

53. John can paint his car in 2 hours, and Joe can paint a similar one in 3 hours. How long would it take them to paint one together?

54. Phoebe and Phyllis run a typing service. Phoebe can type a paper in 50 minutes or the two can type it together in 30 minutes. How long would it take Phyllis to type the same length paper alone?

55. Kel, Del, and Mel are painters. Kel can paint a room in 5 hours, Del in 4 hours, and Mel in 6 hours. One day they all start to work on a room, but after an hour Del and Mel are called to another job and Kel finishes the room. How long will it take him?

56. Bud and Karen went fishing. At the end of the day they compared catches. Together they had caught 40 fish. If Bud had caught two more and Karen had caught two fewer, they would have caught the same number of fish. What was the size of each catch?

57. The Barretts took an automobile trip to Canada. The gas averaged $1.25 per gallon. They spent $60 per night on motels and $40 per day on meals. They averaged 120 miles per day and got 30 miles per gallon on gas. If the total cost of the trip was $2625, how many days were they gone and how many miles did they travel? (They were gone the same number of days as nights.)

58. A 4-inch wide picture frame surrounds a picture 2 inches longer than it is wide. If the area of the frame is 208 square inches, what are the dimensions of the picture? (Hint: Find the difference between the area of the picture and the area formed by the outside dimensions of the frame.)

59. A rectangular swimming pool is surrounded by a walk. The area of the pool is 323 square feet and the outside dimensions of the walk are 20 feet by 22 feet. How wide is the walk? (Quadratic)

60. A lab assistant needed 20 ounces of a 10% solution of sulphuric acid. If he had 20 ounces of a 15% solution, how much must he draw off and carefully replace with distilled water in order to reduce it to a 10% solution?

61. Hank and Hossein open a print shop. They receive a job which Hank can do in 12 hours or Hossein in 14. They start to work on it together, but after 3 hours, Hossein has to stop to finish another job. Hank works alone for an hour, when he is called out on an estimate. Hossein comes back and finishes the original job alone. How long will it take him to finish?

62. A reservoir can be filled by an inlet pipe in 24 hours and emptied by an outlet pipe in 28 hours. The foreman starts to fill the reservoir but he forgets to close the outlet pipe. Six hours later he remembers and closes the outlet. How long does it take altogether to fill the reservoir?

63. At the Indianapolis 500, Carter and Daniels were participants. Daniels' motor blew after 240 miles and Carter was out after 270 miles. If Carter's average rate was 20 mph more than Daniels' and their total time was 3 hours, how fast was each averaging?

64. The pilot of the Western Airlines flight from Los Angeles to Honolulu announced en route that the plane was flying at a certain airspeed with a west wind blowing at 30 mph. However, after flying for 3 hours, the wind decreased to 20 mph (airspeed constant).

The plane arrived in Honolulu 2 hours later. If the distance from Los Angeles to Honolulu is 2500 miles, what did the pilot announce was the airspeed of the plane?

65. Mr. Monte spent 3% of his salary on property taxes, 4% went for health insurance, and he put 10% in the bank. If what remained of his yearly salary was then $30,500 per year, what was his yearly salary?

ANSWERS

1. 25, 86

2. 12, 36

3. 17, 18, 19

4. 23

5. 81, 83, 85

6. 22, 24, 26

7. 2 feet

8. 27

9. 43

10. 83

11. $28,000, $22,000

12. $30.60

13. $12,000, $14,000, $26,000

14. $900 for first house, $1200 for second, $1200 for third.

15. 130 hours for Heinrich
140 hours for Wolfgang

16. $4800, $4000

17. 2 liters

18. $3\frac{3}{5}$, $2\frac{2}{5}$ quarts

19. $\frac{2}{9}$, $1\frac{7}{9}$ quarts

20. $2\frac{6}{7}$ quarts

21. $1\frac{2}{3}$, $3\frac{1}{3}$ quarts

22. $\frac{1}{2}$ hour

23. 9 hours

24. 2 miles

25. 24 minutes, 2 miles, 8 miles

26. 2 hours

27. 90 men, 30 women

28. 80 lbs coconut macaroons,
20 lbs maple dates

29. 40 mph

30. 600 mph

31. 4 feet from fulcrum
on Janie's side

32. 10 pounds

33. 5 feet from 300 pounds
6 feet from 250 pounds

34. $5\frac{1}{2}$ feet

35. 80 pounds

36. $116\frac{2}{3}$ pounds

37. 200 children's, 4000 adult's

38. 3 nickels, 5 dimes, 2 quarters

39. 200 nickels, 600 pennies

40. 10, 40, 2, 2

41. 20 dollars, 100 fifty cent pieces, 200 dimes, 607 pennies

42. 22, 32

43. 12, 24

44. 3 hours, 6 miles

45. 20, 40

46. 93, 96

47. 11, 33

48. 15 by 3 feet

49. 42 feet

50. 35°, 70°, 75°

51. 50 feet, 30 feet

52. 8 by 12 feet

53. $\frac{6}{5}$ hours

54. 1 hour, 15 minutes (or 75 minutes)

55. $1\frac{11}{12}$ hours

56. 18, 22

57. 25 days, 3000 miles

58. 8 by 10 inches

59. $1\frac{1}{2}$ feet

60. $6\frac{2}{3}$ ounces

61. $6\frac{1}{3}$ hours

62. $23\frac{1}{7}$ hours

63. 160 mph Daniels' rate, 180 mph Carter's rate

64. 526 mph

65. $50,000 yearly salary